简明英汉-汉英
岩土工程词汇手册

赵明华　郑　刚　刘汉龙　主编
邹新军　张　玲　刘小平　参编

中国建筑工业出版社

图书在版编目(CIP)数据

简明英汉-汉英岩土工程词汇手册/赵明华,郑刚,刘汉龙主编. —北京:中国建筑工业出版社,2008
 ISBN 978-7-112-10035-4

Ⅰ.简… Ⅱ.①赵…②郑…③刘… Ⅲ.岩土工程-词汇-手册-英、汉 Ⅳ.TU4-61

中国版本图书馆 CIP 数据核字(2008)第 048745 号

简明英汉-汉英岩土工程词汇手册
赵明华 郑 刚 刘汉龙 主编
邹新军 张 玲 刘小平 参编
*
中国建筑工业出版社出版、发行(北京西郊百万庄)
各地新华书店、建筑书店经销
北京天成排版公司制版
北京市密东印刷有限公司印刷

开本:787×1092毫米 1/48 印张:11⅙ 字数:460千字
2008 年 7 月第一版 2008 年 7 月第一次印刷
印数:1—3000 册 定价:**29.00元**
 ISBN 978-7-112-10035-4
 (16838)
版权所有 翻印必究
如有印装质量问题,可寄本社退换
(邮政编码 100037)

本书是一本便携式工具书,收录岩土工程专业词汇近8000条。为了便于读者掌握和节约篇幅,本手册对词汇的释义尽可能选用最习惯用法。

本手册的特色是便携、英汉和汉英合在一本书中。此方式充分考虑读者需求,装在口袋中可以随身携带,随时翻阅。既能由英语查汉语,也能由汉语查英语。

本书适合岩土工程、土木工程、建筑工程、水利工程、交通工程、地下工程、工程勘察等专业设计施工人员使用,也可供大中专学校师生参考。

* * *

责任编辑:郭　栋
责任设计:董建平
责任校对:梁珊珊　兰曼利

目 录

英汉部分	1
A	1
B	16
C	27
D	53
E	68
F	80
G	92
H	101
I	106
J	116
K	118
L	120
M	131
N	143
O	148
P	154
Q	179
R	180
S	199

T	245
U	257
V	262
W	266
Y	272
Z	272
汉英部分	274
A	274
B	275
C	288
D	304
E	329
F	329
G	338
H	350
J	361
K	381
L	390
M	400
N	405
O	411
P	412
Q	420
R	427
S	432

T	455
W	468
X	473
Y	481
Z	502

英汉部分

A

abamurus 挡土墙，扶壁，支墩块
abandoned channel 废河道，牛轭湖
ability to self-compact 自压实性
ablation 消融，磨蚀
ablation breccia 垮塌角砾岩
ablation moraine 消融碛
abnormal water level 异常水位
abra 岩洞，地穴
abraded platform 浪成台地
abrasion 磨蚀
abrasion of blown sand 风砂磨蚀
abrasion platform (terrace) 浪蚀台地，海蚀台地
abrasion resistance test 抗磨损试验
abrasion table-land 剥蚀台地
abrasion terrace 海蚀台地
abrupt slope 陡坡
absolute altitude 绝对标高
absolute chronology 绝对年代，绝对年代学
absolute contour 绝对等高线
absolute control point 绝对控制点
absolute coordinate 绝对坐标
absolute density 绝对密度
absolute elevation 绝对高程(标高，海拔)
absolute humidity 绝对湿度
absolute value 绝对值
absorbed layer 吸附层
absorbed water 吸着水，吸附水

absorbing well 吸水井，渗水井
absorptance 吸收率
absorption limit 吸附界限
absorption loss 吸附损失
absorptive capacity 吸附能力
abutment 桥台，坝肩
abutment deformation 坝座变形
abutment pressure 拱座压力，桥台压力
abutment wall 翼墙，拱座墙
abysmal area(region) 深海区
abysmal clay 深海黏土
abysmal deposit 深海沉积
abyssal facies 深海相
abyssal ooze 深海软泥
abyssal red earth 深海红土
abyssal rock 深成岩
abyssal sea 深海
abyssal sediment 深海沉积物
accelerated cement 快凝水泥
accelerated consolidation 加速固结
accelerated creep of rock 岩石加速蠕变
acceleration of gravity 重力加速度
accelerator 催化剂
acceptable limit 容许极限
acceptance of grout 吸浆量
acceptance rate 吸浆率
acceptance test 验收试验
access road 便道
access tunnel 交通隧洞，进出隧洞
acclimation 适应环境(水土、气候等)
accretion 结核，增生
accumulated error 累积误差
accumulation area 堆积区，冰川
accumulation curve 累积曲线

accumulation terrace 堆积阶地
accumulational platform 堆积台地
accuracy of measurement 量测精度
acicular crystal 针状晶体
acid soil 酸性土
acidic content 含酸量
acidite 酸性岩,硅质岩
acidity and alkalinity test 酸碱度试验
acidity test 酸度试验
acidizing 酸化
acoustic blanking 消声作用(打桩)
acoustic box 消声罩(打桩)
acoustic depth sounding 回声测深
acoustic emission monitoring 声发射监测
acoustic emission of rock 岩石声发射
acoustic method 声学法
acoustic piezometer 声发射测压计
acoustic prospecting 声学探测
acoustic sounder 回声测深仪
acoustic strain gauge 声发射应变计
acoustic(al) log 声法测井
acoustic(al) velocity logging 声速测井
acrylamide grouting 丙凝灌浆
acting head 有效水头
action of gravity 重力作用
activated bentonite 活性膨润土
active agent 活化剂
active clay 活性黏土
active containment alternative 主动抑压法(环保)
active earth pressure 主动土压力
active fault 活断层,活性断层
active force 主动力,有效力
active fracture 活断裂
active layer 活性层(地基土),冻融活性层

active measured value 实测值
active pile 主动桩
active Rankine zone 主动朗肯区
active shear stress 主动剪应力
active slide area 主动滑动区
active soil 活性土
active stabilizer 活性加固剂
active state of plastic equilibrium
　　　主动塑性平衡状态
active surface of sliding 主动滑动面
active water 活性水(有侵蚀性)
active zone 活动层(冻土，膨胀土)
activity 活性
activity index 活动性指数，活动指数，活性指数
activity ratio 活性比
actual breaking stress 真实断裂应力
actual quantity 实际工程量
actual standard 现行标准
actual stress 实在应力
adamic earth 红黏土
additional borehole 补充孔
additional clauses 附加条款
additional load 附加荷载
additional stress 附加应力
additional survey 补充勘测
additive 掺合料，添加剂
adfreezing force 冻附力
adfreezing strength 冻附强度
adhere 黏附
adhesion 黏着作用，黏着力，附着力
adhesion action 胶结作用
adhesion character 黏附性
adhesion factor 黏着系数
adhesive 接合剂

adhesive weight 附着重力
adiabatic compression 绝热压缩
adit (水平)坑道,横坑(道),平硐
adjacent panels 相邻槽段
adjacent(adjoining) rock 围岩
adjusted elevation 平差高程
admissible load 容许荷载
admissible strain 容许应变
admissible stress 容许应力
admixture stabilization 拌合加固法
admixtures 掺合物
adobe 土坯
adsorbed layer 吸附层
adsorbed water 吸附水
adsorbent pressure 吸附压力
adsorption layer 吸附层
adularia 冰长石
adustion 烘干
advanced casing 跟进套管
advanced grouting 超前灌浆
adverse geologic phenomena 不良地质现象
aeolian deposit(soil) 风积土
aeolian erosion 风蚀
aeolian sediment 风积物
aeolianite 风成岩
aerated water 饱气水,碳酸水
aeration zone 通气带
aeration zone water 包气带水
aerial geology 航空地质
aerial sediment 大气沉积物
aerometer 气体比重计
aeronautical chart 飞行图
affinity of water 亲水性
after effect 残余塑流

after-working 后效
afterbreak 地面沉陷破坏
aftershock 余震
aged clay 老黏土
agglomeration 凝聚作用
aggregate 团粒，骨料
aggregate for concrete 混凝土骨料
aggregate structure 团粒结构，聚合体结构
aggregation 集合体
aggregation force 聚合力
aggressive water 侵蚀性水，侵进水
aging 老化
aging effect 时效（强夯）
agitation 搅拌
agitator 搅拌器
agrology 土壤学
A-horizon 淋溶层
air back valve 空气止回阀
air bleed valve 排气阀，放气阀
air compressor 空气压缩机
air content 含气率
air drilling 压缩空气洗井钻进
air driven hammer 风洞锤
air flush drilling 空气钻探
air free water 无气水
air gun 气枪（海洋勘探）
air in soil 土中气
air lift 气举
air motor 风动发动机
air motor drill 风动钻机
air permeability 透气性
air permeability test 透气性试验
air pollution 空气污染
air ram 风动锤

air release valve 排气阀
air space ratio 气隙比
air track drill 履带式风动钻机
air void ratio 气隙比
air-blown concrete 喷射混凝土
air-curtain method 气幕法
air-dried soil 风干土
air-entraining chemical compound 掺气剂
air-entry permeameter 掺气渗透仪
airfield classification system 机场分类法
airlift pump 气举泵,气压泵
air-lock 气闸
air-monitoring device 大气监测仪
airphoto interpretation 航空照片判读
air-supported structures 充气结构物
air-water surface 气水结合面
aleurolite(siltite) 粉砂岩
alimentation 补给
alimentation area 补给区
A-line A线(塑性图)
alkali(ne) soil 碱性土
alkaline water 碱性水
alkalinity 碱度,碱化程度
all rock breakwater 堆石防波堤
all round pressure 周围压力
allomorph 同质同晶
allophane(allophanite) 水铝英石
allowable amplitude 容许振幅
allowable bearing capacity 容许承载力
allowable bearing capacity of foundation soil
　　　地基容许承载力
allowable bearing pressure 容许承载力
allowable bearing value 容许承载力
allowable bond stress 容许黏着应力

allowable deflection 容许偏斜
allowable deformation of subsoil 地基允许变形值
allowable displacement 容许位移
allowable error 容许误差
allowable grade of slope 边坡坡度容许值
allowable gross bearing pressure 总容许荷载
allowable load 允许荷载，容许载重
allowable net bearing pressure 净容许承载力
allowable pile bearing load 容许单桩荷载
allowable pressure 容许压力
allowable relative deformation 容许相对变形
allowable settlement 容许沉降
allowable slenderness ratio 容许长细比
allowable soil pressure 容许土压力
allowable stress 容许应力
allowable thickness of residual frost layer
　　　容许残留冻土层厚度
allowable value of deformation 允许变形值
allowable value of width-height of foundation steps
　　　基础台阶宽高比的容许值
allowable vibration acceleration 容许振动加速度
allowable working load 容许使用荷载，容许
　　　工作压力
alloy drill bit 合金钻头
alluvial 冲积土
alluvial apron 冲积裙
alluvial deposit 冲积物
alluvial fan 冲积扇
alluvial flat 河漫滩
alluvial formation 冲击地层
alluvial ground water 冲积层地下水
alluvial plain 冲积平原
alluvial soil 冲积土，冲积土壤
alluvial terrace 冲积阶地

alluviation 冲积
alluvion 冲积地
alluvium 冲积层，冲积土，淤积层
alluvium grouting 冲击层灌浆
alluvium period 第四纪
alpha method α法(桩工)
alteration 蚀变，变更，交替，改建
altered rock 蚀变岩石
alternate material 代用材料
alternating beds 互层
alternating load 交变荷载
alternating strain 交变应变
alternating stress 交变应力
alternative scheme 替代方案
alum clay 明矾黏土，铝土
aluminum octahedron 铝氧八面体
amargosite 膨润土
ambient pressure 周围压力
ammersooite 阿米水云母
ammonium soil 铵质土
amorphous 非晶质
amorphous substance 无定形物质
amount of compression 压缩量
amount of deflection 挠度
amount of inclination 倾角，倾斜角
amount of leakage 渗漏量
amplification factor 放大因素
amplitude log 声幅测井
amplitude magnification factor 振幅放大因素
amplitude of vibration 振幅
amplitude ratio 振幅比
analysis on interaction of superstructure and foundation 上部结构和基础的共同作用分析
anatectic earthquake 深源地震

anchor 锚定物
anchor bar 锚杆
anchor block 锚块
anchor bolt 锚栓
anchor grout 锚固浆液
anchor grouting 锚固灌浆
anchor head 锚头
anchor hole 锚孔
anchor pile 锚桩
anchor plate 锚板
anchor rod(tie) 锚杆
anchor rope 锚索
anchor slab wall 锚定板墙
anchorage 锚定作用，地锚
anchorage bond strength 锚固粘结强度
anchorage bulkhead 锚定岸壁，锚定挡墙
anchorage cable 锚索
anchorage device 锚定装置
anchorage length 锚固应力损失
anchorage point 锚固点
anchored earth 锚定土
anchored in rock piles 嵌岩柱
anchored pier 锚固墩
anchored sheet pile(piling) wall 锚定板桩墙
anchored tied-back shoring system 后拉锚定系统
anchoring 锚固
anchoring force 锚固力
anchoring plug 锚塞
ancient rock slide 古滑坡
andesite 安山岩
andesite basalt 安山玄武岩
andesite-tuff 安山凝灰岩
anemoarenyte 风成砂岩
angle of base friction 基底摩擦角

angle of bedding 地层倾斜角
angle of bevel 坡口角度
angle of contact 接触角(弯液面)
angle of dilatancy 剪胀角
angle of external friction 外摩擦角
angle of incidence 入射角
angle of inclination 倾角
angle of internal friction 内摩擦角
angle of nature repose 自然休止角
angle of obliquity 倾角(应力)
angle of repose 休止角
angle of rupture 破裂角,土坡破裂角
angle of shear (shearing resistance, shearing strength) 剪切角
angle of slide 滑动角
angle of slope 坡角
angle of true internal friction 真内摩擦角
angle of wall friction 墙摩擦角
angle pile 斜桩
angstrom 埃(Å)
angular displacement 角位移
angular distortion 角挠曲
angular grain 角粒
angular pebbles 角砾
angularity chart 棱角度对比图
angularity factor 棱角度因素
anion 阴离子
anion adsorption 阴离子吸附
anion exchange 阴离子交换
anisotropic consolidation 各向不等压固结
anisotropic hardening 各向不等压硬化
anisotropic soil 各向异性土
anisotropy 各向异性
anisotropy of rock mass 岩体的各向异性

annual precipitation 年降雨量
annual runoff 年径流量
annular bit 环形钻头
anomite 褐云母
anorthose 斜长石
anthroplithic age 石器时代
anticenter of earthquake (anti-epicenter, anti-epicentrum) 震中对点,反震中
anticline 背斜
anti-corrosion protection coat 防腐涂层
anti-flocculation 反絮凝作用
antifoam 阻沫剂(钻探)
anti-scour wall 防冲墙
anti-seep diaphragm 防渗膜,截渗墙
anti-slide pile 抗滑桩
anti-vacuum 反压力
anti-vibrating stability 抗震稳定性
anvil 砧
A-parameter 孔隙压力系数 A
apatite 磷灰石
aperiodic damping 非周期阻尼
aperiodic motion 非周期运动
aperture 空口,孔径,小孔,开度
aperture of rock fissure 岩石裂隙张开度
aphanitic texture 隐晶结构(火成岩)
aphanocrystalline texture 隐晶结构(沉积岩)
API method API 法(桩工)
apparent angle of internal friction 表观内摩擦角
apparent coefficient of friction 表观摩擦系数
apparent cohesion 表观黏聚力
apparent density 堆积密度,表观密度
apparent preconsolidation pressure 表观先期固结压力

apparent shearing strength 表观抗剪强度
apparent specific gravity 表观比重，体比重
apparent velocity 表观流速
apparent viscosity 表观黏度
applied fore 外加力，作用力
applied stress 作用应力
approach pit 导坑(基础托换)
apron 护坦，冰川前砂砾层
apron crib 护坦笼框
apron plain 冰碛平原，冰川沉积平原
apron slab 护坦板
apron stone 护脚石
apron wall 前护墙
aquatic chemistry 水化学
aqueoglacial deposit 冰水沉积
aqueous soil 水成土，饱水土
aquiclude 阻水层，隔水层
aquifer 含水层
aquifuge 不透水层
aquitard 弱透水层
arbitrary assumption 任意假定
arch action 拱作用
archaeozoic era 太古代
arching 拱作用
architectonic geology 构造地质学
area of pumping depression 抽水下降面积
area ratio 面积比
arenaceous sediment 砂岩，砂质沉积物
arenite(arenyte) 砂屑岩
areometer 比重计
argillaceous sandstone 泥质砂岩
argillaceous sediment 泥质沉积
argillaceous shale 泥质页岩
argillaceous slate 泥质板岩

argillite 矽质黏土岩，粘板岩
arid region 干旱区
arkose 长石砂岩
arkose quartzite 长石石英岩
armor 人工防冲铺盖
arrangement of piles 桩的布置
arrangement of piles in rank form 桩的行列式排列
arrival time 到达历时（物探）
artesian aquifer 承压含水层
artesian flow 承压水流，自流
artesian pressure head 承压水头，自流水压力
artesian water 承压水
artesian well 自流井
artificial fill(soil) 人工填土
artificial foundation(ground) 人工地基
artificial recharge of ground water 地下水人工补给
artificially improved soil 人工加固土
asbestos 石棉
aseismatic design 抗震设计
aseismic region 无震区
ashstone 凝灰岩
askanite 蒙脱石
asperity 粗糙度
asphalt 地沥青
asphaltic grouting 沥青灌浆
asphaltic sealing 沥青堵漏
aspirator 流水抽气器
associative flow rule 相适应流动法则
assumed load 计算荷载，假定荷载
at rest pressure 静止土压力
atmosphere 大气圈
atmospheric pressure 大气压力

atmospheric pressure method 真空预压法(软基加固)
atmospheric water 大气水
atomic bonds 原子键
attenuation 衰减
Atterberg limits 阿太堡界限
attitude 产状
audio equipment 声频仪
auger bit 螺旋钻头
auger boring 螺旋钻探
auger drill 螺旋钻
auger pile 螺旋桩
autographic odometer 自录计程仪
automated header 自动调节集管
automated grout plant 自动制浆站
automated grouting equipment 自动灌浆设备
automated injection system 自动注入系统
automatic ram pile driver 自动冲锤打桩机
avalanche 雪崩，山崩，崩坍
avalanche dam 崩塌坝(地质)
avalanche prevention works 防塌工程
aven 落水洞，竖井
average(grain)diameter 平均粒径
average contact pressure 平均接触压力
average modulus 平均模量
average rock condition 中等岩石条件
average settlement 平均沉降
average stress 平均应力
average tensile strength 平均抗拉强度
average thickness method 平均厚度法
axial allowable bearing capacity of pile
    桩的轴向容许承载力
axial bearing capacity 轴向承载力
axial bearing capacity of drill caisson embedded

in bedrock 嵌岩管柱轴向承载力
axial bearing capacity of piles 桩的轴向承载力
axial compression 轴向压缩
axial compressive force 轴向压力
axial deformation 轴向变形
axial extension test 轴向拉伸试验
axial load 轴向荷载,轴心荷载
axial load test 轴向荷载试验
axial strain 轴向应变
axial stress 轴向应力
axial symmetry 轴对称
axially symmetric consolidation 轴对称固结
axisymmetric stress 轴对称应力

B

back analysis 反演分析
back digger 反向铲
back drain 墙背排水设施
back fill 回填土
back fill density 回填密度
back filling 回填
back grouting 二次灌浆
back of levee 堤防背水面
back pressure 反压力
back scatter densimeter 同位素反射密度仪
back slope 后坡
backfill density 回填密度
backfill grouting 回填灌浆
backfill tamper 回填夯
backflow connection 回流管
backflow valve 回流阀
backhoe 反铲,反铲挖土机

backosmotic pressure　反渗压力
backshore terrace　滨后阶地
backswamp　漫滩沼泽
backward erosion　向源浸蚀
backwater　回水
bag dam　土袋埝坝
bahada　山麓冲积平原
bailer　抽泥筒
balance cone　平衡锥(稠度试验)
ball and socket joint　球窝接头
ballast　道碴，压载
ballasting of mattress　柴排压沉
balloon densimeter　囊式密度计
balloon volumeter　囊式体积仪
band shaped prefabricated drain　带状预制排水板
banded(bandy)clay　带状黏土
bank caving　坍岸
bank drill　杆钻，架式凿岩机
bank gravel　岸砾石
bank materials　岸积物，土堤坝建筑材料
bank measure　填方量
bank protection　护岸
bank protection work　护岸工程
bank settlement　河岸沉陷
bank stabilization　堤岸加固，岸坡稳定处理
bank strengthening　堤岸加固
bank-run gravel　采石坑砾石
banquette　弃土堆
bar gravel　沙洲砾石
bar screen　格栅
barite　重晶石
barium sulphate test　硫酸钡试验
Barlin stone　巴林石
barrel　岩芯管，圆管

barrier 潜堰，截水墙
barrier effect 遮帘作用(桩工)
barrier lake 堰塞湖
Barron's consolidation theory 巴隆固结理论
Baryta feldspar 钡长石
barytes 重晶石
basal heave 基坑底隆起
basal instability 坑底不稳定性
basalt 玄武岩
basanite 碧玄岩
base circle 坡底圆
base course 道路基层
base exchange 盐基交换
base exchange capacity 盐基交换容量
base failure of slope 土坡基底破坏
base line 基准线
base of a quay wall 岸壁基础
base of excavation 开挖基线
base of levee 堤基
base of slope 坡脚
base rock 基岩
base tilt factor 基底倾斜因素
basement 基底，基础，地下室
basement wall 地下室墙
basement water proofing 地下室防水
baseplate 底板
basic intensity 基本烈度
basic value of bearing capacity 承载力基本值
batching equipment 配料设备
bathyal deposit 半深海沉积
bathymetric line 等深线
batter brace 斜撑
batter drainage 斜坡排水
batter leader pile driver 斜导架式打桩机，打

斜桩机
batter pile 斜桩
Baume gravity 波美比重
Baume hydrometer 波美比重计
Baume scale 波美比重标，波美比重计，波美标度
bay delta 海湾三角洲
bayou(lake) 牛轭湖
beach deposit 滩沉积(海、湖)
beach fill 海岸淤填土
beach sand 海滩砂
beam foundation 梁式基础
beam on elastic foundation 弹性地基梁
beam on elastic foundation analogy
　　　弹性地基梁比拟法
beam pitman 锚筋，拉杆
bearing capacity 承载能力
bearing capacity factor 承载力因数，承载力系数
bearing capacity of foundation 地基承载力
bearing capacity of single pile 单桩承载力
bearing course 承压垫层
bearing failure 承压破坏
bearing force of soil 土的承压力
bearing graph 反应曲线(桩工)
bearing pad 承压垫板
bearing pile 支承桩
bearing pressure 承载压力
bearing ratio 承载比
bearing stratum 持力层
bearing test 静力载荷试验
bearing value 承载力
bed rock 基岩
bedded formation 层状地层
bedded rock 成层岩石，层状岩石

bedded structure 层状构造
bedding 层理，垫料，打底，铺垫
bedding(course) material 垫层材料
bedding concrete 垫层混凝土
bedding course 三合土垫层，垫层，褥垫
bedding cushion 垫层
bedding error 端垫误差(三轴试验)
bedding mortar 垫层砂浆
bedding plane 层面，垫层面，层理面
bedding slip 层面滑动，顺层滑动
bedplate 底座，底板
bed-plate foundation 底板式基础
bedrock 基石
bedrock acceleration 基岩加速度
bedrock map 基岩地质图
beetle head 送桩锤
behaviour 性状
belaying pin 套索桩
belled pier 扩底墩
belled pile 扩底桩
belled shaft 扩底墩
belled-out cylindric(al) pile 大头圆柱桩
belling bucket 扩底挖斗
bench mark 水准点
bench stoping 台阶式开采
benched foundation 台阶式基础
bending moment 弯矩
bending stress 弯曲应力
bentonite 膨润土
bentonite cement grout 膨润土水泥浆
bentonite grouting 膨润土灌浆
bentonite mud(grout) 膨润土泥浆
bentonite slurry 膨润土浆液
berm 戗台，护坡道

berth 停泊处，泊位
beta method β法(桩工)
bevel sheet pile 斜角板桩
biaxial state of stress 双轴应力状态
biaxial tensile test 双轴抗拉试验
bifurcation theory 分叉理论
bilinear model 双线性模型
bimodal curve 双峰形粒径曲线
bin effect 仓效应
binder course 结合层(路工)
Bingham fluid 滨汉流体
Bingham model 滨(宾)汉模型
Bingham substance 滨汉体
biochemical rock 生物化学岩
biogenic deposit 生物沉积
biolith 生物岩
biological weathering 生物风化
bioslime 生物软泥
Biot consolidation theory 比奥固结理论
biotine 钙长石
biotite 黑云母
Bishop and Morgenstern slope stability analysis method 毕肖普与摩根斯坦法(滑坡分析)
Bishop's simplified method of slices
　　　　　　毕肖普简化条分法
bit 钻头，比特
bit density 位密度
bit reamer 扩孔钻，扩孔器
bitumen grouting 沥青灌浆
bituminous clay 沥青黏土
black cotton soil 黑棉土(印度等地)
black soil 黑土
Blaine fineness 布莱恩细度
blanket 铺盖

blanket course 透水铺盖层(地基加固)
blanket grouting 铺盖灌浆
blast densification 爆破振密
blast furnace slag 高炉炉渣
blasting 爆破
blasting compaction method 爆炸挤密法
blasting consolidation 爆破加固
blasting-expand pile 爆扩桩
bleed 泌水,析水
bleed path 泌水路径
bleed test 吸水试验
bleeder well 减压井,排水井
blended soil 混杂土
blind catch basin 截留盲井
blind drain 盲沟
blind drainage 盲沟排水
blind subdrain 盲沟
blinding 淤塞,基础垫层,填封
blinding concrete 混凝土护层(基础开挖)
blinding material 填塞材料,透水材料
B-line B线(塑性图)
block anchorage 块体锚固(块锚法)
block diagram 框图,方块图
block failure 整体破坏(桩工)
block foundation 实体式基础
block sample 块状试样
block shear test 块体剪切试验
block stone 块石
block theory 块体理论
blow count 击数
blow count of SPT 标准贯入击数
blow of dynamic sounding 动力触探锤击
blow of penetration test 贯入试验锤击
blow-in 涌入

blown sand 风成砂
blown tip pile 爆扩桩
blows per-min 每分钟击数
blow-up 鼓起(路工)
blue print 蓝图
body force 体积力,质量力,体力
body measurement method 实体测定法
body stress 体应力,内应力
body wave 体波
bog 沼泽
bog muck 沼泽腐殖土
bog peat 沼泽泥炭土
boiling 砂沸,翻浆
bolt 锚杆,螺栓
bolting 锚固
bond 胶结,键,结合,粘结力
bond clay 胶结黏土
bond strength 键力,粘结强度
bonded stone 拉结石
bonding 胶合
bonding agent 胶粘剂
bonding force 键力,结合力
book clay 带状黏土
bookhouse fabric 书堆组构
border effect 周边效应
border pile 边桩,边缘桩,边界桩
border stone 镶边石
bore casing 套管
bore core 岩芯
bore hole 钻孔
bore hole seismic logging 地震测井
bore log 钻孔柱状图
bore plug 钻孔土样
bore rod 钻杆

bored cast-in-place pile 钻孔灌注桩
bored pile 钻孔桩
borehole caliper 钻孔规
borehole camera 钻孔照相机
borehole cleaning 清孔
borehole core 钻孔岩芯
borehole direct shear device 钻孔直剪仪
borehole extensometer 钻孔引伸仪
borehole gravimeter 钻孔重力仪
borehole jack 钻孔千斤顶
borehole(boring)log 钻孔记录，钻孔柱状图
borehole periscope 钻孔潜望镜
borehole pressure recovery test 钻孔压力恢复试验
borehole record 钻孔记录，钻孔柱状图
borehole sample 钻孔土样
borehole scanner 钻孔扫描器
bore-hole seismic logging 地震测井
borehole shear apparatus 钻孔剪切仪
borehole specimen 钻孔土样
borehole surveying 钻孔勘探
borehole TV 钻孔电视
borehole wall 孔壁
borer 钻工，钻机，钻头
borescope 钻孔检查显示器(桩工)
boring 钻探
boring bit 钻头
boring casting 钻探套管
boring frame 钻架
boring log 钻孔记录，钻孔柱状图
boring machine 钻机
boring rig 钻探机具
boring rod 钻杆
Borros point 波罗杆(沉降观测)
borrow 取土，采料

borrow area 取土场，采料场
borrow exploration 料场勘探
borrow pit 取土坑
Boston blue clay 波士顿蓝黏土
bottom heave 基底隆起
bottom pressure 底压力（隧道）
boulder 漂石，漂砾，圆(砾)石，块石
boulder clay 漂石黏土
boulder foundation 大块卵石填沙基础
bound water 吸附水，结合水
boundary conditions 边界条件
boundary layer 边界层
Bourdon gauge 布尔登压力计
Boussinesq theory 布辛涅斯克理论
bowl scraper 铲运机
bowl-shape settlement 碟形沉降
box caisson 沉箱
box foundation 箱形基础
box pile 箱形钢桩
box shear apparatus 盒式剪切仪
B-parameter 孔隙压力系数 B
braced excavation 支撑开挖
braced sheeting 支撑挡板
bracing 支撑
brackish deposit 半咸沉积
brackish water 半咸水
Brazilian test 巴西试验，劈裂试验
break stone 碎石
breakdown pressure 破碎压力
breaking down test 断裂(破坏)试验
breaking edge 顶板岩石的崩落线，崩落线
breaking load 破坏荷载
breaking strain 断裂应变
breaking stress 断裂应力

breast wall 胸墙
breccia 角砾岩
brick foundation 砖基础
bridge 钻孔堵塞(孔壁塌落造成)
bridge pier 桥墩
briefly frozen soil 暂冻土
brine 卤水
brine pond 卤水池
brine storage 卤水存储
brittle clay 脆性黏土
brittle failure 脆性破坏
brittle fracture 脆性破裂
brittle rock 脆性岩石
brittleness index 脆性指数
broadly graded soils 宽级配土
broken course 断层
broken gravel 碎砾石
broken sliding surface 不连续滑动面
broken stone 碎石
broken stone base course 碎石垫层
bromhead ring shear 环剪
brooming 木桩碎裂
Brownian movement 布朗运动
brucite 水镁石
bucket auger 勺钻
buckling 压曲，纵向弯曲
buckling load 压曲荷载
buffer 减振器
bulb pile 球根桩，扩端桩
bulk compressibility 体积压缩性
bulk density 体密度
bulk modulus 体积模量（M_b）
bulk specific gravity 体比重，假比重
bulk unit weight 体积单位重量，体容量

bulkhead(wall) 岸壁，挡土墙
bulking 体胀
bulldozer 推土机
buoyancy 浮托力
buoyancy density 浸没密度
buoyancy raft 浮筏基础
buoyant foundation 筏形基础
buoyant pile 浮桩，摩擦桩
buoyant unit weight 浮重度
Bureau Public Road Clarification System
　　美国公路局土的分类法［BPRCS］
burette 滴定管
Burger model 伯格模型
buried structure 埋入式结构物
buried terrace 埋藏阶地
burst strength 胀破强度
burst tearing test 胀破试验
bust pump 钻孔泵
butt 桩头
buttress retaining wall 支墩式挡墙
buttressed wall 扶垛墙
by-pass seepage 绕渗

C

cable-hung clamshell bucket 绳索悬挂瓣式抓斗
cable-tool drill(ing) 钢绳冲击钻(探)
cage of reinforcement 钢筋笼
Cainozoic era 新生代
caisson 沉箱
caisson foundation 沉箱基础
caisson pile 管柱桩，沉箱桩，钻孔灌注桩
calcareous concretion 钙质结核

calcareous ooze 钙质软泥
calcareous soil 石灰质土
calcareous spar 方解石
calcilutite(calcilutyte) 灰泥岩
calcite 方解石
calcite dolomite 方解白云石
calcite marble 方解大理石
calcspar 方解石
calculated self-weight collapse
　　　　　计算自重性湿陷量(Δ_{zs})
calculous soil 砾质土
calibration 标定
calibration curve 标定曲线
calibration test 标定试验
California Bearing Ratio(CBR)
　　　　　加州承载比(美国)
caliper log 测径记录
calyx drill 弯状取芯钻
calyx drill holes 弯状钻孔
camber 坝顶超高
Cambrian period 寒武纪
Cambridge model 剑桥模型
Cambridge sampler 海底取样器
Camkometer 剑桥旁压仪
camouflet 地下空间，地下爆炸
canalization 渠道化
cannel shale 烛煤页岩
cantilever footing 悬臂式基础
cantilever retaining wall 悬臂式挡土墙
cantilever sheet pile wall(cantilever sheet piling)
　　　悬臂式板桩墙
cap model 盖帽模型
cap of pile 桩帽，桩台
capacity of well 井出水量

capblock 桩帽
capillarimeter 毛细管仪
capillarity 毛细管作用
capillary film 吸附水膜
capillary fringe 毛细管水上升边缘
capillary height 毛细管水上升高度
capillary interstice 毛细管间隙
capillary lift test 毛细管水上升高度试验
capillary migration 毛细管水移动
capillary potential 毛细管水势
capillary pressure 毛细管水压力
capillary saturation 毛细管水饱和
capillary tension 毛细管水张力
capillary water 毛细管水
capillary zone 毛细管水带
capped yield model 盖帽屈服模型
capsule-type pressure gauge 囊式压力计
carbonaceous sediment 碳质沉积
carbonatite 碳酸岩
Carboniferous period 石炭纪
cardboard drain 纸板排水
cardboard wick 排水纸板
Carlson stress meter 卡尔逊应力计
case hardening 表面硬化
case history 工程实录
case study 实例研究
cased bore pile 套管钻孔桩
cased borehole 套管钻孔
cased concrete pile 套管混凝土桩
cased pile 套管桩
casing 套管
casing collar 套管接触
casing drive adapter 套管传动联接器
casing head gas 套管头气体

casing shoe 套管靴
casing tubes 套管
Cassagrande dot 卡萨格兰德孔隙压力测头
Cassagrande liquid limit apparatus
　　卡萨格兰德液限仪
Cassagrande's soil classification
　　卡萨格兰德土分类法
cast-in situ pile(cast-in-place pile) 灌注桩
cast-in-place diaphragm wall 就地灌注地下连续墙
cast-in-situ pile by excavation 挖孔桩
cat and can 履带式铲运拖拉机
catchment area 汇水面积
categorization of geotechnical projects
　　岩土工程分级
cathodic protection 阴极保护
cation 阳离子
cation exchange 阳离子交换
cation exchange activity 阳离子交换活动性
cation exchange capacity [CEC] 阳离子交换量
caustic lime mud 灰质软泥
cave deposit 洞穴沉积
cave in 冒顶，下陷，塌落
cavern 洞穴
cavern rock 多孔岩石，溶洞岩石
caving 坍落
cavity 天然洞穴
cavity expansion theory 扩孔理论
CBR value 加州承载力比值(美国)
cell action factor 盒作用因数
cell pressure 室压，盒压，腔压
cell quarry wall 格式岸壁
cell test 荷兰式三轴试验
cellular bulkhead 格式岸壁 [钢板桩]

cellular cofferdam 格形围堰
cellular core wall 箱格形心墙
cellular retaining wall 格形挡土墙
cellular texture 网格组织
cellulosic fibers 纤维素纤维
celsian feldspar 钡长石
cement deep mixing composite foundation
　　深层搅拌桩复合地基
cement grouting(injection) 水泥灌浆
cement mortar 水泥砂浆
cement soil stabilization 水泥土加固
cement stabilization 水泥加固
cementation 胶结作用
cemented soil 胶结土
cementing agent 胶结剂
cement-stabilized soil 水泥加固土
cement-treated-soil grout 水泥土浆
Cenozoic era 新生代
center line 中心线
center peg 中心桩
centering method 中线法(尾矿坝)
central earthquake 中心震
central load 中心荷载,轴心荷载
centrifugal autoclaved concrete pile
　　离心蒸压混凝土桩
centrifugal concrete pile 离心混凝土桩
centrifugal force 离心力
centrifugal model test 离心模型试验
centrifugal prestressed concrete pile
　　离心预应力混凝土桩
centrifugal pump 离心泵
centrifugal reinforced concrete pile
　　离心钢筋混凝土桩
centrifugally-cast concrete driven pile

离心法制混凝土打入桩
centrifuge 离心机
centrifuge modeling 离心机模型
centrifuge moisture equivalent 离心含水当量
centrifuge test 离心试验
centroid 重心，形心
ceramic filter 陶瓷滤头
ceramics 陶管
cerruti's solution 色卢铁解答
chalk 白垩
chalk marl 白垩泥灰岩
chalky clay 白垩黏土
chance sample 随机样品
channel 渠道
chat 碎石
chaser 送桩器
check dam 谷坊
check hole 检查孔
chellean age 旧石器时代(舍利时代)
chemical additive 化学添加剂
chemical board drain 化学板排水，排水化学板
chemical bonding 化学胶合法
chemical churning pile 旋喷桩
chemical decomposition 化学分解
chemical deposit 化学沉积
chemical grouting 化学灌浆
chemical injection process 化学注浆法
chemical interaction 化学相互作用
chemical precipitation 化学沉积
chemical reagents 化学试剂
chemical stability 化学稳定性
chemical stabilization 化学加固
chemical waste disposal 化学废料处理
chemical weathering 化学风化

chemise 土堤岸护墙,土堤岸护面
chernozem soil 黑钙土
chimney drain 竖向截水体(坝工)
chisel 冲击钻头,凿
chisel bit 冲击式一字形钻头
chlorite 绿泥石
choking up 淤塞,淤填
chopping bit 冲击钻头
C-horizon 母质层
chronological time scale 地层时序表
chunk sample 块状试样
churn drill 旋冲钻
circle of failure 破坏圆
circle of stress 应力圆
circular arc analysis 圆弧分析法
circular cofferdam 圆形围堰
circular failure analysis 圆弧破坏分析
circular foundation 圆形基础
circular sliding surface 圆弧滑动面
circular slip surface 圆弧滑动面
circular soil cutter 切土环刀
circumferential stress 环向应力
clasolite 碎屑岩
class of hardness 硬度分类
classification 分类
classification of soil 土的分类
classification test 分类试验
clastic deposit 碎屑沉积
clay 黏土
clay binder 黏土胶结物
clay blanket 黏土铺盖(水工),黏土护层(隧道)
clay chunks 黏土块
clay content 黏粒含量
clay core 黏土心墙

clay cutter	切土筒
clay fraction	黏粒粒组
clay grouting	黏粒灌浆
clay lens	黏土透镜体
clay mineral	黏土矿物
clay onion-skin bond	黏粒葱皮粘结
clay pan	隔水黏土层，隔水盘
clay parings	黏土夹层
clay platelet	黏土片晶
clay puddle	黏土膏
clay rock	黏土岩
clay sandstone	黏土砂岩
clay shale	黏土质页岩
clay slate	黏土板岩
clay vein	黏土脉
clay-cement grouting	黏土水泥灌浆
clayey loam	黏壤土
clayey sand	黏质砂土
clayey silt	黏质粉土
clayey soil	黏性土
clayite	高岭石
clayslide	黏土滑坡
claystone	黏土岩
clean sand	纯砂
clean-out auger	清孔钻
clean-out tools	清孔机具
cleavage	劈理，节理
cleaving stone	板岩
cliff debris	坡积物
C-line	C 线（塑性图）
clinometer	测斜仪
clod test	土块实验
clogging	淤塞，堵塞作用
close pile	密排桩

closed conduit 暗管
closed system 封闭系统
club-footed pile 扩底桩
cluster 桩群
cluster bent 群桩排架
cluster structure 团粒结构
clustered piles 桩束(群),集桩,群桩
coagulant 凝聚剂
coagulation 凝聚作用
coagulator 促凝剂
coarse aggregate 粗骨料
coarse fraction content 粗粒组合量
coarse-grained soil 粗粒土
coarse gravel 粗砾
coarse sand 粗砂
coarse screen 粗筛
coarse silt 粗粉土
coarse waste 粗岩屑
coarsely granular particle size 粗颗粒粒度
coarseness of grading 颗粒粒度,级配粒度
coast deposit 海岸沉积
coast sand 海岸砂
coast terrace 海岸阶地
coastal ground water 海岸地下水
coated gravel 含壳砾石
coated pile 涂面桩
coaxial nozzles 同轴喷嘴
coaxiality 共轴性
cobble 卵石,碎石
cobbly soil 粗砾质石
cobstone 大卵石
codes of practice 行业规范
coefficient between layers 层间系数,管涌比
coefficient index 压缩指数

coefficient of active earth pressure　主动土压力系数
coefficient of adhesion　黏着系数
coefficient of attenuation　衰减系数
coefficient of collapsibility　湿陷系数
coefficient of compaction　压实系数
coefficient of compressibility(compression)
　　　压缩系数
coefficient of compressibility of soil
　　　土的压缩系数
coefficient of consolidation　固结系数，压实系数
coefficient of consolidation for radial flow
　　　径向固结系数
coefficient of creep　蠕［徐］变系数
coefficient of curvature　曲率系数
coefficient of deformation due to leaching
　　　溶滤变形系数
coefficient of dynamic viscosity　动力黏滞系数
coefficient of dynamic(al) subgrade reaction
　　　地基反力动力系数
coefficient of earth pressure at rest
　　　静止土压力系数
coefficient of elastic non-uniform compression
　　　弹性非均匀压缩系数
coefficient of elastic shear　弹性剪切系数
coefficient of elastic uniform compression
　　　弹性均匀压缩系数
coefficient of foundation ditch's rebound
　　　基坑回弹系数
coefficient of friction　摩擦系数
coefficient of horizontal consolidation
　　　水平固结系数
coefficient of horizontal pile reaction
　　　桩的水平向反力系数
coefficient of horizontal pressure　侧压系数

coefficient of imperviousness 不透水系数
coefficient of infiltration 渗透系数
coefficient of internal friction 内摩擦系数
coefficient of kinematic viscosity
　　　运动黏滞系数
coefficient of linear expansion 线性膨胀系数
coefficient of passive earth pressure
　　　被动土压力系数
coefficient of permeability 渗透系数
coefficient of resilience 回弹系数
coefficient of restitution 恢复系数
coefficient of rigidity 刚度系数
coefficient of secondary consolidation
　　　次固结系数
coefficient of seismic effect 地震影响系数
coefficient of self-weight collapsibility
　　　自重湿陷系数
coefficient of shrinkage 收缩系数
coefficient of shrinkage of soil
　　　土的收缩系数
coefficient of sorting 分选系数
coefficient of stiffness 刚度系数
coefficient of subgrade reaction
　　　地基反力系数，路基反力系数
coefficient of swelling 膨胀系数
coefficient of thaw-subsidence 融沉系数
coefficient of thermoosmotic transmission
　　　热渗系数
coefficient of transmissibility 导水系数
coefficient of uniformity 不均匀系数
coefficient of vertical consolidation
　　　竖向固结系数
coefficient of vertical pile reaction
　　　桩的竖向反力系数

coefficient of viscosity 黏滞系数
coefficient of volume compressibility
　　体积压缩系数
coefficient of volume compressibility of soil
　　土的体积压缩系数
coefficient of water saturation 饱和吸水率
coefficient of weathering 风化系数
coercibility 可压缩(可凝)性
cofferdam 围堰
cofferdam-bracing 围堰支撑
cohesiometer 黏度计
cohesion 黏聚力，内聚力，凝聚力
cohesionless soil 无黏性土
cohesive sediment 黏性泥沙
cohesive soil 黏性土
cohesive soil foundation 粘结土地基
collapse 坍陷，湿陷
collapse earthquake 塌陷地震
collapse load 极限荷载，破坏荷载，压溃荷载
collapse settlement 坍陷，湿陷
collapse state 破坏状态
collapsibility 湿陷性
collapsibility grading index 分级湿陷量
collapsibility of loess 黄土湿陷性
collapsibility test of loess 黄土湿陷试验
collapsible filter 自落填缝反滤层
collapsible loess 湿陷性黄土
collapsible settlement 湿陷量
collapsible soil 湿陷性土
collapsible subsoil 湿陷性地基
collapsing soil 崩塌土
collector drain 集水沟
collector well 集水井
colloid fraction 胶粒粒组

colloidal grout　胶体悬液
colloidal particle　胶粒
colloidal suspension　胶体悬液
colluvial clay(soil)　崩积土
colluvium　崩积层
colonnade foundation　管柱基础
colonnade foundation process　管柱基础施工法
column pile　端承桩，柱桩
columnar section　柱状图
columnar structure　柱状结构
combined bolting and shotcrete　喷锚支护
combined end-bearing and friction pile
　　　端承摩擦桩
combined footing　联合基础，联合桩基，复式
　　　基础
combined water　结合水
compact rock　致密岩石
compacted column　挤密桩
compacted depth　压实深度，夯实深度
compacted expanded base concrete pile　夯实
compacted fill(compacted soil)　压实填土，夯
　　　实填土
compacted lift　压实分层厚度，夯实分层厚度
compactibility　压实性，可夯实性
compacting factor　压实系数，夯实系数
compacting machinery　碾压机械
compaction(compacting)　压实，夯实
compaction by compaction　置换挤密
compaction by rolling　碾压
compaction by tamping　夯实法
compaction by vibrating roller　振动碾压法
compaction by vibration　振动压实
compaction by watering　注水压实
compaction curve　击实曲线，压实曲线

compaction device 击实仪
compaction effort 击实功,压实功
compaction factor 压实系数
compaction grouting 挤密灌浆
compaction grouting method 挤密喷浆法
compaction in layers 分层压实
compaction mold 击实筒
compaction of deep bed 深层压实
compaction of soil 土的压实
compaction parameter 压实参数
compaction pile 挤密桩
compaction pile method 挤密桩法
compaction plant 碾压设备
compaction sand pile 挤密砂桩
compaction test(compacting test)
　　击实试验,压实试验
compactive effort 压实力,压实效应
compactness 密实度
compactness of soil 土的密实度
compactor 压实机
compactor pass 压实遍数
compatibility equation 相容方程
compensated foundation 补偿式基础
competent degree of rock 坚固度
complete sliding 整体滑动
completely penetrated well 完整井
complex modulus 复模量
complexity of site 场地复杂性
composite foundation(ground) 复合地基
composite geotextile 组合型土工织物
composite ground with lime-fly ash columns
　　二灰土桩复合地基
composite pile 混合桩复合管
composite sliding(slip) surface 复合滑动面

composite slope 复式边坡
compound alluvial fan 复合冲积扇
compound reinforcement 复式加筋
comprehensive engineering geological map 综合工程地质图
compress-crushed zone 挤压破碎带
compressed-air caisson 气压沉箱
compressed-air hammer 压缩空气锤
compressed-air method 气压法(隧道)
compressed-air tamper 压缩空气夯
compressibility 压缩性
compressibility foundation 压缩性地基
compressibility of soil 土的压缩性
compressibility ratio 压缩比
compressible soil 压缩性土
compression 压力，压缩
compression and back expansion 剪缩回胀性
compression capacity 受压承载力
compression curve 压缩曲线
compression curve of soil 土的压缩曲线
compression deformation 压缩变形
compression index 压缩指数
compression of soil 土的压缩性
compression ratio 压缩比
compression strength 抗压强度
compression test 压缩试验
compression wave 压缩波，P 波
compression zone 压缩区
compressive deformation 压缩变形
compressive(compression) strain 压应变
compressive strength 抗压强度
compressive(compression) stress 压应力
concentrated force 集中力
concentrated load 集中荷载，点荷载

concentration factor 集中因素(应力)
concentration index 集中指数
concentric load 中心荷载(轴心荷载)
concrete cushion 混凝土垫层
concrete deadman 混凝土锚桩，混凝土锚墙
concrete face rockfill dam 混凝土面板堆石坝
concrete foundation 混凝土基础
concrete gravity platform 混凝土重力式钻采平台
concrete key trench 混凝土截水槽[土坝]
concrete pile 混凝土桩
concrete rubble 混凝土碎石
concrete sheet piling 混凝土板桩
concrete tubular pile 混凝土管桩
concrete vibra column 振动混凝土柱(桩)
concrete-timber pile 混凝土木桩
concreting 浇灌混凝土
condensation water 凝结水
condition of shear strength 剪切强度条件
condition of strength 强度条件
conductivity 导热率，电导率
conductivity of medium 介质传导率
conductor pipe 导管
conduit 管道
conduit jacking 顶管法
cone bearing test 圆锥承重试验
cone foundation 锥形基础
cone of depression 下降漏斗(地下水位)
cone of influence 影响漏斗(地下水)
cone of water-table depression 地下水位漏斗线
cone penetration impact test 锥击试验
cone penetration test (CPT) 静力触探试验，圆锥触探试验
cone penetrometer 圆锥触探仪
cone penetrometer for liquid limit test

锥式液限仪
cone resistance 圆锥探头阻力
confindence level 置信度
confindence limit 置信界限
confined aquifer 层间含水层
confined compression test 侧限压缩试验
confined compressive strength 侧限抗压强度
confined water 承压水，层间水
confined water head 承压水头
confining pressure 围压
confirmatory measurement 核实测量
conformable stratum 整合地层
conglomerate 砾岩
conical penetrometer 圆锥触探器
conical pile 锥形桩
conjugate stress 共轭应力
connate water 原生水，封存水
consequent landslide 顺层滑坡
consistency 稠度
consistency index 稠度指数
consistency limit 稠度界限
consistency of soil 土的稠度
consistency test 稠度试验
consolidated anisotropically drained test
  各向不等压固结排水试验
consolidated anisotropically undrained test
  各向不等压固结不排水试验
consolidated drained test 固结排水试验
consolidated drained triaxial test
  固结排水三轴试验
consolidated isotropically undrained test
  各向等压固结不排水试验
consolidated quick(direct) shear test
  固结快剪试验

consolidated slow shear test 固结慢剪试验
consolidated undrained compression test
　　　固结不排水压缩试验
consolidated undrained shear test 固结不排水剪切试验
consolidated undrained test 固结不排水试验
consolidated undrained triaxial test
　　　固结不排水三轴试验
consolidated-drained direct shear test
　　　固结排水直剪试验，慢剪试验
consolidated-drained triaxial compression test
　　　固结不排水三轴压缩试验
consolidating pile 加固桩
consolidation 固结，加固
consolidation apparatus 固结仪
consolidation by dewatering 排水固结法
consolidation by electroosmosis 电渗加固
consolidation by preloading and drainage
　　　预压排水固结
consolidation by the vacuum method 真空固结
consolidation by vibration
　　　振动固结，振动压密
consolidation curve 固结(渗压)曲线
consolidation deformation 固结变形
consolidation failure 固结破坏
consolidation grouting 固结灌浆
consolidation line 固结曲线
consolidation of collapsing soil
　　　湿陷性黄土地基加固
consolidation of expanding soil
　　　膨胀土地基加固
consolidation of karst(carst)
　　　喀斯特地基加固，岩溶地基加固
consolidation of liquefiable sandy soil

液化砂土地基加固

consolidation of soft subsoil 软基加固
consolidation of soil 土的固结
consolidation of soil mass 土体固结
consolidation pressure 固结压力
consolidation ratio 固结比
consolidation sedimentation 固结沉淀作用
consolidation settlement 固结沉降，压密沉降
consolidation test 固结试验，压缩试验
consolidation test under constant loading rate
　　等速加荷固结试验
consolidation test under constant rate of strain
　　等应变速率固结试验
consolidation test under controlled gradient
　　控制比降固结试验
consolidation theory 固结理论，压密理论
consolidation theory of soil 土的固结理论
consolidation under K_0 condition K_0 固结
consolidometer 固结仪
constant coefficient method of lateral loading analysis of single pile 单桩横向受力分析的常系数法
constant gradient test 等梯度固结试验
constant head permeability test 常水头渗透试验
constant head permeameter 常水头渗透仪
constant pressure device 恒压装置
constant rate of penetration test 等速贯入试验（桩工）
constant rate of uplift test 等速上拔试验（桩工）
constant volume test 恒体积试验
constant-force-amplitude excitation
　　恒力幅激振
constitutive equations 本构方程

constitutive law 本构定律,本构定理
constitutive law of soil 土的本构定律
constitutive model of soil 土的本构模型
constitutive relation 本构关系
constitutive relation of rock (constitutive relationship of rock mass) 岩石本构关系
constitutive surface 构造面
constrained diameter (grain size) 限制粒径
constrained modulus 压缩模量
constraint equation 约束方程
constraint of motion 运动约束
constraint reacting force 约束反(作用)力
construction terraces 堆积阶地
contact area 接触面积
contact erosion 接触面冲蚀
contact grouting 接触灌浆
contact pressure 基底压力,接触压力
contact scouring 接触面冲蚀
contact stress transducer 接触应力传感器
contact-altered rock 触变岩
contaminated soil 受污染土
contemporary glaciation 现代冰川作用
continental glacier 大陆冰川
continental sedimentation 大陆沉积,陆相沉积
continental shelf 大陆架
continental slope 大陆坡
continual loading test 连续加荷固结试验
continuity condition 连续条件
continuity equation of seepage 渗流连续方程
continuous beam 连续梁
continuous concrete (slurry) wall 地下连续墙
continuous coring 连续取岩芯
continuous creep 连续蠕变
continuous footing 连续底座,连续基础

continuous foundation 连续基础
continuous grading 连续级配
continuous heat test 连续加热试验
continuous medium 连续介质
continuous piled wall 连续桩成墙
continuous sampling 连续取样
continuous velocity log 连续速度测井
continuously loaded oedometer 连续加荷固结仪
continuum 连续介质
continuum mechanics 连续介质力学
contour line 等值线
contour map 等高线图，等值线图
contour map of groundwater 地下水等水位线图
contractile skin 结合膜，收缩表面
contraction fissure 收缩裂缝
contraction-joint grouting 收缩缝灌浆
contractive soil 收缩性土
control of underground water 地下水控制
controlled blasting technique 收控爆破技术
controlled-strain test 压变控制试验
controlled-stress test 应力控制试验
conventional triaxial test for rocks
   岩石常规三轴试验
conventional triaxial testing machine for rocks
   岩石常规三轴试验机岩芯钻探
convexity 凸度
coombe 冲沟
core 岩芯，心墙
core barrel 岩芯钻筒
core bit 岩芯钻头
core catcher 采芯器
core cutter 岩芯切取筒
core drill 岩芯钻
core drill rig 岩芯钻机

core earth dam 心墙土坝
core hole 岩芯钻孔
core of rock 岩芯
core recovery 岩芯采取率,岩芯回收率
core sampler 岩芯取样器
core trench 心墙截水槽
core wall 心墙
core wall of dam 不透水的坝心墙
core wall type rockfill dam 心墙堆石坝
core-leaving method 旁侧导坑法(隧道)
coring 钻取岩芯
coring bit 岩芯钻头
corner-points method 角点法(应力计算)
corrasion 腐蚀,侵蚀
correlation coefficient 相关系数
corrosion 刻蚀作用
corrosion rate 腐蚀率
corrosion resistance 耐蚀性
corrosive water 侵蚀性水
corundum 刚玉
Coulomb criterion 库仑准则
Coulomb-Navier strength theory
　　　库伦-纳维强度理论
Coulomb's earth pressure 库仑土压力
Coulomb's earth pressure theory
　　　库仑土压力理论
Coulomb's equation for shear strength
　　　库仑抗剪强度公式
Coulomb's law 库仑定律
Coulomb's soil-failure prism
　　　库仑土破坏楔体
counter-bore 扩孔钻
counterfort retaining wall 扶壁式挡土墙
counterpoising 压坡

counterweight fill 压重填土
country rock 围岩，母岩
coupled vibration 耦合振动
covalent bond 共价键
coverage 覆盖范围，涂层
covering layer 覆盖层
crack 裂缝
crack gauge 测缝计
crack grouting 裂缝灌浆
crack growth 裂纹扩展
crack reflection 裂缝反射（路工）
crack water 裂隙水
cracked tension zone 张裂区
crater 夯坑（强夯）
creep 蠕［徐］变，蠕动
creep behaviour test 蠕［徐］变性状试验
creep damage 蠕［徐］变损伤
creep deformation 蠕［徐］变损伤变形
creep function 蠕［徐］变函数
creep limit 蠕［徐］变极限
creep load 蠕［徐］变荷载
creep motion 蠕［徐］变运动
creep rate 蠕［徐］变速率
creep rupture 蠕［徐］变破坏
creep settlement 蠕［徐］变沉降
creep strain 蠕［徐］变应变
creep strength 蠕［徐］变强度
creep test 蠕［徐］变试验
creeping rock mass 蠕滑断层
creeping waste 蠕动坡积物
creep-path length 爬径长度（渗流）
Cretaceous period 白垩纪
crevice-water 裂隙水
crib 格排

crib retaining wall 框格式挡土墙
critical circle 临界圆
critical damping 临界阻尼
critical density 临界密度
critical depth of excavation 开挖临界深度
critical edge pressure 临塑荷载
critical exit gradient 临界逸出梯度
critical floatation gradient 临界浮动梯度(砂土的)
critical frequency 临界频率
critical height of slope (土坡)临界高度
critical hydraulic gradient 临界水力梯度
critical length of pile 临界桩长
critical load of subsoil 地基临界荷载
critical load(ing) 临界荷载
critical number of blows of standard penetration test 临界标准贯入击数
critical particle size 临界粒度
critical slope 临界坡度
critical state energy theory 临界物态能量理论
critical state locus 临界物态轨迹
critical state model 临界物态模型
critical state soil mechanics 临界物态土力学
critical state strength 临界物态强度
critical surface 临界面
critical void ratio 临界孔隙比
cross bedding 交错层面
cross bit 十字钻头
cross mouthed drill 十字形钻
cross section 截面
cross section line stake 横断面桩
cross section profile 横断面图
cross sensitivity 正交测斜灵敏度
cross-anisotropic soil 交叉各向异性土
cross-arm settlement gauge 十字臂沉降管

crossed strip foundation　十字交叉条形基础
cross-hole method(shooting)　跨孔法
cross-plane permeability　横面渗透性
cross-stratification　交错层理
crown plate　顶板
cruciform clay cutter　十字形黏土冲切器
crumbling rock　崩解岩石
crumb-structure　团粒结构
crumpling　褶皱作用
crushed aggregate　碎料
crushed boulder　碎漂石
crushed broken stone　粗碎石
crushed gravel　碎砾石
crushed rock(stone)　碎石
crushed stone soil foundation　碎石土地基
crushed-run aggregate　破碎料
crushing strength of grains　颗粒压碎强度
crust　地壳，硬表层
crust breccia　断层角砾石
crystal edge　晶棱
crystal imperfection　晶体不完整
crystal lattice　晶格
crystal structure　晶格结构
crystal(lization) water　结晶水
crystalline rock　结晶岩，结晶岩石
cube strength　立方体强度
cubic dilatation(cubical expansion)
　　　　体(积)膨胀
cubical triaxial test apparatus　真三轴试验仪
Culmann construction　库尔曼图解法(土压力)
Culmann line　库尔曼线
Culmann's method　库尔曼法
culvert　涵洞
cumularsharolith　团粒

cumulative curve 累积曲线（颗粒分析）
cumulose soil 腐殖土
curbstone 拦石
cure 固化
curtain grouting 帷幕灌浆
curve fitting 曲线拟合
curve of equal settlement 等沉陷曲线
curve stake 曲线桩
cushion 垫层，桩垫
cushion pile 缓冲桩，垫桩
cushion stress factor 垫层应力系数
cut 挖方，路堑
cut diameter 分割粒径
cut off wall 截水墙
cut slope 路堑边坡，切坡
cut-and-cover 随挖随填，挖方填方
cut-and-fill 挖方填方，边冲边淤
cut-off elevation 断桩高程
cutting 路堑
cutting edge 刀脚
cutting jet 切割射流
cutting resistance 抗切断强度
cutting ring 环刀
cuttings of boring 钻屑
cycle of vibration 振动循环
cyclic degradation 周期衰化
cyclic heat test 周期加热试验
cyclic load test 周期载荷试验
cyclic loading(load) 周期荷载
cyclic shear resistance 周期抗剪强度
cyclic simple shear test 周期单剪试验
cyclic strain softening 周期应变软化
cyclic stress 周期应力
cyclic triaxial test 周期加荷三轴试验

cyclone pump 旋流泵
cyclotheric sedimentation 旋回沉积作用
cylinder 沉井，圆柱体
cylinder pile 大直径桩
cylinder strength 圆柱体强度
cylindric(al) pile 圆筒桩

D

daily drilling report 钻孔日报表
daily grouting report 灌浆日报表
dam failure 坝失事，溃坝
dam foundation 坝基
dam foundation grouting 大坝基础灌浆
dam heel 坝踵，坝的上游坡脚
dam site investigation 坝址勘查
dam site survey 坝址测量
dam toe 坝趾，坝的下游坡脚
damage index 损坏指数
damming 筑坝，壅水
damp proof course 防潮层
damp proofing 防潮
damped vibration 有阻尼振动
damping 阻尼
damping factor 阻尼因素
damping matrix 阻尼矩阵
damping ratio 阻尼比
damping ratio of soil 土的阻尼比
Darcy's law 达西定律
dashpot 阻尼器
datum plane 基准面
dauk 砂质黏土，亚黏土
dead lime 失效石灰

dead load 恒载
dead shoring 固墙竖撑
deadman 深埋锚定桩
debris 岩屑
debris flow 泥石流
debris laden stream 含大量沙砾的河流
debris slide 岩屑滑动
decay period 衰减周期
decay time 衰减时间
decayed rock 风化岩层
decline of underground water level
　　地下水位下降
decompaction 振松
decomposed granite 风化花岗石
decomposed rock 风化岩石
decomposition 分解作用
decompressed rock 减压岩石
decompression 卸荷，沉箱减压
decompression and recompression loop 卸压再压环
decompression curve 卸荷曲线
decompression modulus 卸荷模量
decontamination 排除污染
decreasing axial pressure fracture test
　　轴向减压断裂试验
decrementation 分级卸荷
deep blasting 深孔爆破，深层爆炸
deep boring 深钻孔，深层钻探
deep compaction 深层压实
deep compaction by vibration 振动深层压密
deep consolidation 深层加固
deep cut 深开挖
deep densification 深层加密
deep excavation 深开挖

deep foundation 深基础
deep jet mixing pile machine 深层混合搅拌桩机器
deep mixing method 深层搅拌法
deep ocean ooze 深海软泥
deep percolation 深层渗透
deep pumped well 深泵井
deep sea deposit 深海沉积物
deep settlement 深层沉降
deep settlement gauge 深层沉降仪
deep slide 深层滑坡
deep soil stabilization 深层土加固
deep sounding apparatus 测深仪
deep sounding test 深层触探试验
deep well method 深井法
deep-lime-mixing method 深层石灰搅拌法
deep-sea sediment 深海沉积物
deep-seated grouting 深层灌浆
deep-well pump 深井泵
deferred grouting 延迟灌浆
deflation 风蚀
deflected pile 偏位桩
deflection curve 挠度曲线
deflection inclinometer 测斜仪
deflectometer 挠度计
deflocculating agent 反絮凝剂,分散剂
deflocculation 反絮凝作用
deformation 变形
deformation conditions 变形条件
deformation control 变形控制
deformation due to shrinkage of soil
　　　土的收缩变形量
deformation modulus of rock 岩石变形模量
deformation modulus of rock mass

岩体变形模量
deformation of expansion of soil
　　土的膨胀变形量
deformation of rock mass　岩体变形模量
deformation of soil mass　土体变形
deformation parameters test　变形参数试验
deformation resolution　变形分辨率(量测仪器)
deformation softening　应变软化
deformation stress　变形应力
deformational energy of rock　岩石的变形能
deformational pressure of surrounding rock
　　围岩变形压力
deformeter　变形仪
defrosting　解冻
degradation　剥蚀，衰化
degradation index　衰化指数
degradation parameter　衰化参数
degree Baume　波美度
degree of compaction　压实度，密实度
degree of consistency　稠度
degree of consolidation　固结度
degree of dryness　干度
degree of expansion　膨胀度
degree of fineness　细度
degree of freedom　自由度
degree of liquefaction　液化度
degree of remoulding　重塑度
degree of rock weathering　岩石风化程度
degree of saturation　饱和度
degree of sensitivity　灵敏度
degree of shrinkage　收缩度
degree of vaccum　真空度
degree of weathering　风化程度
degree-days　度-日(冻土用单位)

dehydration test 脱水试验
delayed compression 次压缩
delayed consolidation 次固结
delayed elasticity 弹性后效
delayed settlement 延时沉降，次固结沉降
Delft continuous sampler 德佛特连续取样器
detonation 起爆
delta 三角洲
deltic cross bedding 三角洲交错层
dense sand 密砂
densification 加密
densification by explosion 爆炸加密法
densification by sand pile 挤密砂桩
densimeter 密度计
density 密度
density bottle 比重瓶
density gauge 密度测定器
density index 密度指数
density of dry soil 干土密度
density of probe （同位素）密度探测仪
density of saturated soil 饱和土密度
density of soil 土的密度
density of solid particles 土粒密度
density of submerged soil 浸没土的密度
density of water 水的密度
density test 密度试验
denudation 剥蚀作用
deposit 沉积，沉积物
depletion coefficient 疏干系数
deposit in situ 原地沉积
depression-cone 降水漏斗
depressive coefficient of soil 土壤抗陷系数
depth and width corrections for bearing capacity value 承载力值的深宽修正

depth creep 深层蠕动
depth effect of pile group 群桩深度效应
depth factor 深度因素
depth of embedment 埋深
depth of footing 大放脚埋深
depth of foundation 基础埋深
depth of frost penetration 冻结深度
depth of influence 影响深度
depth of overburden 覆盖层厚度
depth of penetration 贯入深度
depth of scour 冲刷深度
depth ratio 埋深比(基础)
depth resistance curve of pile 桩的深度-阻力曲线
derrick stone 巨石块
desaturation 减饱和作用
descent of regional ground water level 区域性地下水位下降
desert soil 荒漠土
desiccation 干燥作用
desiccator 干燥器
design criteria 设计准则
design(fortification)intensity 设计(设防)烈度
design lift 设计压实层厚
design load 设计荷载
destressed zone 应力解除区
detachable bit 可拆卸式钻头
detailed exploration 详细勘探
detrimental settlement 有害沉降
detrital sediment 碎屑沉积
detritus 碎石
developing chart of exploratory drift 坑碉展示图
deviation 偏位
deviator stress 偏应力
deviatoric state of stress 偏应力状态

device　设备，仪表
devolution　崩塌
Devonian period　泥盆纪
dewatering　降低地下水位，疏干
dewatering method　降水法
dewatering project　降水工程
dewatering system　降水系统
D-horizon　D层，下伏岩土层
diagenesis　成岩作用
diagonal bedding　斜层理
diagonal fault　斜断层
diagonal joint　斜节理
dial gauge(clock gauge)　量表，千分表
diameter distribution　粒径分布
diameter of sediment gain　泥沙粒径
diamond crown bit　金刚石钻头
diamond drilling　金刚石钻头钻探
diaphragm　隔板，薄膜
diaphragm pump　隔膜泵
diaphragm wall　截水墙，地下连续墙
diatomaceous earth(diatomite)　硅藻土
diesel pile hammer　柴油打桩锤
differential equation of constitution
　　　本构微分方程
differential equation of continuity
　　　连续性微分方程
differential equilibrium equation of Euler
　　　欧拉平衡微分方程
differential frost heave　不均匀冻胀
differential heave　不均匀隆胀
differential rock bolt extensometer
　　　差动式岩石锚杆引伸仪
differential settlement　沉降差
difficult foundation　难处理地基

difficult ground condition　不良地质条件
diffraction　衍射
diffuse double layer　扩散双电层
diffuse seepage　扩散渗流
diffusion　渗滤
diffusion layer　扩散层
diffusivity　扩散性
dike(dyke, levee)　堤
dilapidation　崩落
dilatable soil　膨胀土
dilatancy　剪胀性
dilatancy equation　剪胀方程
dilatancy of rock　岩石扩容
dilatation　剪胀，松胀
dilatation of soil mass　土体的剪胀
dilatational wave　疏密波，P波
dilation　剪胀，松胀
dilative soil　剪胀性土
dilatometer　膨胀仪
diluvial fan　洪积扇
diluvial soil　洪积土
diluvion(diluvium)　洪积层
dimensional analysis　量纲分析
dimensional ratio　量纲比
diorite　闪长岩
dip (angle)　倾角
dip joint　倾向节理
dip slip fault　倾向滑断层
dipmeter　倾斜仪，测斜仪
dipolar ion　偶极离子
dipolar moment　偶极矩
dipole　偶极子
dip slope　反坡
direct flushing　正循环冲洗

direct shear apparatus 直剪仪
direct shear test 直剪试验
direct simple shear test [DSS-test]
　　　　直接单剪试验
direct stress 正应力
directional drilling 定向钻进
directional jet grouting 定灌喷浆
directional shear cell 定向剪切盒
dirt band 冰川碎石带
disaggregation 分散作用
disaster geology 灾害地质学
discharge 流量
discharge orifice 排水孔，泄流孔
discharge velocity 排水速度，泄流速度
discoloured clay 变色黏土
discontinuity structural plane 不连续构造面
discretization 离散化
disintegrative 崩解性
disk pile 盘底桩
dislocation 断层，位错
dispersed structure 分散结构
dispersant(dispersing agent) 分散剂
dispersion 分散作用
dispersion coefficient 弥散系数，分散系数
dispersity 分散性，分散度
dispersive clay(soil) 分散性黏土
displacement 位移
displacement criterion of slop safety
　　　　边坡安全性的位移判据
displacement fault 平移断层
displacement grouting 排水灌浆
displacement method 挤淤法
displacement of surrounding rock 围岩位移
displacement pile 打入桩，排土桩，挤压桩

displacement pump 活塞泵
displacement ratio 置换率
disposal area 弃土区
dissipation 消散
dissipation test 消散试验(孔隙水压力)
dissolution 溶解
dissolution basin 溶蚀盆地
dissolved salts 溶解盐
distance of settlement influence 沉降影响深度
distilled water 蒸馏水
distortion settlement 瞬时沉降
distortional wave 剪切波，S 波
distressed zone 应力解除区
distributed load 分布荷载
distribution of grain size 颗粒级配，粒径分布
disturbance degree 扰动度
disturbance index 扰动指数
disturbance ratio 扰动比
disturbed soil sample 扰动土样
ditch 明沟
divalent ion 双价离子
diversion tunnel 导流洞
divisional plane 节理面，断层面，劈理面
dolinae(doline) 落水洞，斗淋
dolly 桩垫
dolomite 白云岩
dolomite limestone 白云石灰岩
dolomite marble 白云大理石
dolphin 系缆柱
dolphin type breasting structure 桩式靠船建筑
domain 叠片体，域
domain fabric 粒团组构
dome foundation 穹窿式基础
dosimeter 量筒

dot recording system 点式记录仪
double amplitude 双振幅
double bridge type penetrometer 双桥式触探仪
double core barrel 双层岩芯管
double diaphragm pressure cell (double diaphragm pressure gauge) 双膜式土压力盒
double layer 双电层
double lee 四通
double shear test 双面剪切试验
double tube sampler 双层取样器
double wall cofferdam 双排板桩围堰
double-acting steam hammer 双动汽锤
double-layer soil foundation 双层地基
double-packer grouting 双塞灌浆
double-specimen oedometer test 双样固结试验
double-well sheet pile cofferdam 双壁板桩围堰
dowel pile 钉桩
downdrag 下拉荷载（桩工）
down-hole hammer 潜孔锤
down-hole method 下孔法
down-hole percussive drill 冲击式潜孔钻机
down-hole shooting 下孔法
down-hole TV camera 孔下点式录像机
downstream method 下游法（尾矿坝）
dozer 推土机
drag 摩阻力
drag boat 挖泥船
drag force 曳阻力
dragline 拉铲
drain 排水管，排水沟
drain pile 排水砂桩，砂井
drain trench 排水沟
drainage 排水
drainage basin of groundwater 地下水流域

drainage blanket 排水垫层
drainage by consolidation 固结排水
drainage by desiccation 疏干
drainage by electro-osmosis 电渗排水
drainage by gravity 重力排水
drainage by surcharge 加载排水
drainage by well point 井点排水
drainage characteristics 排渗特性
drainage crossings 交叉排水渠(管)
drainage filter 排水滤层,排水滤体
drainage layer 排水层
drainage loading 排水加荷
drainage method 排水法
drainage of foundation 地基排水
drainage outlet 排水出口
drainage path 排水路径
drainage sump(drainage well) 排水井
drainage system 排水系统
drainage tile 排水瓦管
drained repeated direct shear test
　　排水反复直剪试验
drained shear test 排水剪切试验
drained triaxial test 排水三轴试验
drawdown 水位下降
drawdown curve 水位下降曲线
drawdown ratio 水位下降比
dredge level 疏浚标高
dredge pump 泥浆泵,污水泵
dredged spoils 挖出物
dredger 挖泥船
dredger fill 吹填土
drift 冰碛堆积物,漂积物,冰碛
drift map 覆盖层地质图
drift sand 流砂

drifted material 洪积物
drifted soil 冰碛土
drill bit 钻头
drill hole 钻孔
drill log 钻孔柱状图
drill pipe 钻管，钻杆
drill rig 钻探机具
drillability 可钻性
drilled caisson 钻孔沉井，管柱，沉井
drilled drain hole 排水钻孔
drilled grout hole 灌浆钻孔
drilled maximum diameter 钻孔最大直径
drilled pier 钻孔墩，钻孔桩
drilled shaft 钻孔竖井，钻孔桩
drilling 钻探
drilling and wireline coring 钻探绳索采样系统
drilling diameter 成孔直径
drilling equipments 钻探设备
drilling fluid 钻液
drilling mud 钻孔泥浆
drilling platform 钻采平台，钻井平台
drilling report 钻探报告
drilling rig 钻机
drilling rod 钻杆
drilling time log 钻时记录
drive cast-place-pile 钻孔灌注桩
drive collar 桩箍（桩头铁圈）
drive head 承锤头（触探）
drive pipe 洛阳铲
drive sampler 击入式取土器
drive shoe 桩靴，套管管靴
driven cast-in-place pile 就地灌注桩
driven in-situ pile 打入灌注桩
driven pile 打入桩

driven precast concrete pile　打入预制混凝土桩
driven shell pile　打入包壳桩,打入钢桩
driven test pile　打入试验桩
driven underpinning piles　打入式托换桩
driving band　桩箍(木桩)
driving cap　桩帽
driving depth　打桩深度
driving energy　打桩能量
driving hammer　桩锤
driving helmet　桩帽
driving of sheet piling　打板桩
driving pile abutment　打桩台
driving record　打桩记录
driving resistance　打桩阻力
driving rig　打桩设备
driving shoe　桩靴
driving stress　打桩应力,锤击应力
driving test　打桩试验
drop　落距
drop hammer　落锤
drop pile hammer　打桩落锤
drop sampler　沉落取样器
drop-impact penetrometer　落锤冲击贯入仪
Drucker-Prager criterion　德鲁克-普拉格准则
dry density　干密度
dry drilling　干式钻进
dry oven　烘箱
dry pack　干填(托换工程)
dry permafrost　干永冻层
dry rodding　干捣法(试件制备)
dry screening　干筛
dry strength　干强度
dry tamping　干击法(试件制备)
dry unit weight　干重度

dry-drilled pile　干成孔灌注桩
drying oven　烘箱
drypack　干填砂浆
Ducth cone penetrometer　荷兰锥贯入仪
ductility factor　延性系数
dug cast-in-place pile　挖孔灌注桩
dune　砂丘
duripan　硬盘(土)层
Dutch cell　荷兰三轴仪
Dutch cone penetrometer　荷兰式圆锥触探仪
dyeing　染色法
dyke boring　堤防钻探
dyke defect detecting　堤防隐患探测
dyke maintenance　堤防维修
dynamic analysis　动态分析
dynamic bearing capacity　动承载力
dynamic cone penetrometer　动力圆锥触探仪
dynamic consolidation　强夯法，动力固结
dynamic densification of soils　土的振动压密
dynamic displacement　动态移动
dynamic effective stress method
　　　动力有效应力方法
dynamic elastic modulus　动力弹性模量
dynamic elastic modulus of rock　岩石动弹性模量
dynamic level　动水位
dynamic load　动荷载
dynamic load test of pile　桩的动荷载试验
dynamic magnification factor　动力放大因素
dynamic oedometer　动力固结仪
dynamic penetration test　动力触探试验
dynamic pile driving resistance　打桩动阻力
dynamic pile test　动力测桩
dynamic point resistance　动力探头阻力
dynamic Poisson's ratio of rock

岩石的动泊松比
dynamic pore pressure 动孔隙压力
dynamic properties of rock 岩石的动力特性
dynamic properties of soils 土的动力性质
dynamic property parameter of soil
　　　土的动力性质参数
dynamic reciprocity 动力互换
dynamic replacement 动力置换法
dynamic resistance 动阻力
dynamic response 动力反应
dynamic response approach 动力反应法
dynamic shear modulus of soil 土的动剪切模量
dynamic shear strength 动抗剪强度
dynamic simple shear test 动单剪试验
dynamic sounding 动力触探
dynamic strength of rock 动态岩石强度
dynamic strength of soils 土的动强度
dynamic stress 动应力
dynamic subgrade reaction 动基床反力
dynamic test of pile(s) 桩基的动力试验
dynamic triaxial test 动三轴试验
dynamic viscosity 动力黏滞性
dynamic-pile driving formula 动力打桩公式
dynamite method 爆炸法(排淤、挤密等)
dynamometer 测力计

E

earth 土
earth anchor 土锚杆，土层锚杆
earth auger 螺旋钻，土钻
earth backing 还土

earth beam test 土梁试验
earth borer 土钻
earth cofferdam 土围堰
earth crust 地壳
earth dam 土坝
earth dam ageing 土坝老化
earth dam compaction 土坝压实
earth dam paving 土坝护面
earth drill 土钻
earth embankment 土堤
earth fall 土崩
earth fill 填方
earth flow 泥流
earth foundation 土质地基
earth material 土料
earth membrane 黏土防渗层
earth movement 地壳运动
earth pile 土桩
earth pressure 土压力
earth pressure at rest 静止土压力
earth pressure balanced shield 土压平衡式盾构
earth pressure cell 土压力盒
earth pressure coefficient 土压力系数
earth pressure distribution 土压力分布
earth pressure due to live load 活荷载产生的土压力
earth pressure of loose ground 散体地压
earth slide 土崩，土滑，滑坡
earth slip 土层滑动
earth structure 土工结构物
earth work 土方工程
earthquake 地震
earthquake center 震中
earthquake damage 震害

earthquake dynamic earth pressure 地震动土压力
earthquake effects 地震效应
earthquake engineering 地震工程学
earthquake fault 地震断层
earthquake focus 震源
earthquake hazards 震害
earthquake hypocenter 震源
earthquake intensity 地震烈度
earthquake intensity scale 地震烈度表
earthquake magnitude 地震震级
earthquake proof construction 抗震建筑
earthquake proof foundation 抗震基础
earthquake proof joint 防震缝
earthquake resistant design 抗震设计
earthquake resistant structure 抗震结构
earthquake response spectrum 地震反应谱
earthquake subsidence 震陷
earthquake wave 地震波
earth-rock dam 土石坝
earth-rock excavation 土石方开挖
earthwork 土石方工程
earthwork engineering 土工工程
eboulement 崩坍，滑坡
eccentric load 偏心荷载
eccentrically loaded footing 偏心荷载基础
eccentricity 偏心距
echograph 回声测探仪
edaphology 土壤学
edge disturbance 边缘扰动
edge pressure 边缘压力
edged ring 环刀
edge-to-face flocculation 边对面絮凝
eductor well point 喷射井点
eductor well point system 喷射井点系统

effective angle of internal friction 有效内摩擦角
effective confining pressure 有效周围压力
effective consolidation stress 有效固结应力
effective cross section 有效截面积
effective deformation 有效沉降量，有效体变（强夯）
effective density of soil 土的有效密度
effective grain size(effective diameter) 有效粒径
effective normal stress 有效法向应力
effective opening 有效筛孔
effective over-burden pressure 有效上覆压力
effective porosity 有效孔隙率
effective pressure 有效压力
effective size 有效粒径
effective stress 有效应力
effective stress analysis 有效应力分析
effective stress parameters 有效应力参数
effective stress path 有效应力路径
effective unit weight 有效重度
effective well radius 井的有效半径
efficiency factor 效率系数
efficiency formula 效率公式(群桩)
efflux pump 射流泵
effusive rock 喷出岩
egg box foundation 多格基础
ejector well point 喷射井点
elastic after effect 弹性后效
elastic beam 弹性梁
elastic compression 弹性压缩
elastic constant 弹性常数
elastic deformation 弹性变形
elastic deformational pressure of surrounding rock 围岩弹性变形压力

elastic distortion 弹性畸变
elastic equilibrium 弹性平衡
elastic fill 弹性垫层，非刚性垫层
elastic force 弹性力
elastic formula 弹性公式
elastic foundation 弹性地基
elastic foundation beam analogy method
　　　　弹性地基梁比拟法
elastic foundation beam method 弹性地基梁法
elastic half-space foundation model
　　　　弹性半空间地基模型
elastic half-space theory 弹性半空间理论
elastic heave （土）弹性隆起
elastic hysteresis of rock 岩石的弹性滞后
elastic limit 弹性极限
elastic medium 弹性介质
elastic model of soil 土的弹性模型
elastic modulus 弹性模量
elastic modulus of rock 岩石的弹性模量
elastic modulus of soil 土的弹性模量
elastic semi-infinite body 弹性半无限体
elastic semi-infinite foundation 弹性半无限地基
elastic settlement 弹性沉降
elastic side wall 弹性边墙
elastic solid 弹性体
elastic state of equilibrium 弹性平衡状态
elastic strain 弹性应变
elastic strain energy 弹性应变能
elastic subgrade reaction 弹性基床反力
elastic support 弹性支座
elastic theory 弹性理论
elastic zone 弹性区
elasticity modulus 弹性模量
elastic-plastic analysis 弹塑性分析

elastic-plastic behavior 弹塑性状
elastic-plastic matrix 弹塑性矩阵
elastic-plastic model 弹塑性模型
elastic-plastic solid 弹塑性体
elastic-plasticity of soil 土的弹塑性
elastomer 弹性灌浆料
elastoplastic coupling 弹塑性耦合
elastoplastic range 弹塑性范围
elasto-plasticity theory 弹塑性理论
electric double layer 双电层
electric fluviograph 电测水位计
electric log 电测井
electric profiling 电测剖面法
electric prospecting 电法勘探
electric stabilization 电加固
electrical analog method 电模拟法
electrical analogue for ground water flow
　　　地下水流的电模拟
electrical drainage 电动排水
electrical piezometer 电测孔隙水压力仪
electrical prospecting 电法勘探
electrical sounding 电测深
electrical-drainage 电渗排水
electrochemical effect 电化学效应
electrochemical stabilization 电化学加固
electromagnetic prospecting 电磁勘探
electromagnetic subsurface probing [ESP]
　　　电磁地下测探
electro-osmosis 电渗法
electro-osmotic stabilization 电渗加固
elevation 标高，高程
elevation above sea level 海拔高度
elevation of bore hole 孔口标高
elevation of water 水位

e-log$p$ curve e-log$p$ 压缩曲线
elongation ratio 伸长比
elongation test 拉长试验
elutriation test 淘洗试验
eluvial facies 残积相
eluvial soil 残积土
eluviation 淋溶
eluvium 残积层
embankment 路堤
embankment dam 填筑坝
embankment pile 护堤桩
embedded construction 隐蔽工程，埋设仪器
embedded depth 埋置深度，嵌固深度
embedded length 埋入长度
embedment 埋置，埋置深度
emergency grouting 应急灌浆
emergency measures 应急措施
emergency shaft 应急井，备用井
empirical constant 经验常数
empirical equation(empirical formula) 经验公式
empirical value 经验值
end bearing 端阻力
end bearing pile 端承桩，支承桩
end cap 桩帽
end effect 末端效应
end joint 平接，端接
end resistance of pile 桩端阻力
end restraint effect 端部约束效应
end-bearing capacity of pile 桩端承载能力
end-bearing pile 端承柱
endochronic theory 内时理论
endothermic reaction 吸热反应
enforced settlement 强迫下沉(沉井)
engineered fill 控制回填土

engineering geologic columnar profile
  工程地质柱状图
engineering geologic drilling 工程地质钻探
engineering geologic evaluation 工程地质评价
engineering geologic investigation(exploration)
  工程地质勘探
engineering geologic investigation report
  工程地质勘察报告
engineering geologic map 工程地质图
engineering geologic process 工程地质作用
engineering geologic profile 工程地质剖面图
engineering geologic unit 工程地质单元
engineering geologic zoning 工程地质分区
engineering geological analogy 工程地质类比法
engineering geological condition 工程地质条件
engineering geological evaluation 工程地质评价
engineering geological mapping 工程地质测绘
engineering geological prospecting 工程地质勘探
engineering geologist 工程地质师
engineering geology 工程地质学
engineering geomorphology 工程地貌学
engineering quality index of rock mass
  岩体工程质量指标
engineering soil map 土分布图
enlarged base 扩底(桩工)
enlarging bit 扩眼钻头
enrockment 抛石，堆石
entisol 新成土
entrapped air 封闭空气，截留空气
envelope of failure 破坏包线
envelope of Mohr's circles 摩尔圆包线
environmental geotechnics 环境岩土工程学
environmental hydrogeology 环境水文地质学
eolation 风蚀

eolian deposit 风积物
e-p curve e-p 压缩曲线
epicentral distance 震中距
epicentral intensity 震中烈度
epicentre(epicenter) 震中
epicontinental sedimentation 陆缘沉积,浅海沉积
epoxy bonding 环氧基粘合
epoxy resin 环氧树脂
equation of continuity 连续方程
equilibrium moisture content 平衡含水量
equipotential line 等势线
equipressure lines 等压线
equivalent angle of internal friction
　　　　等效内摩擦角
equivalent beam method 等值梁法
equivalent consolidation pressure 等效固结压力
equivalent damping ratio 等效阻尼比
equivalent diameter 等效粒径
equivalent fluid 等效流体
equivalent footing analogy 等效基础模拟
equivalent grain size 等效粒径
equivalent height of surcharge 超载等效高度
equivalent material 等效材料
equivalent opening size [EOS] 等效孔径
equivalent radius 当量半径
equivalent relative density 当量相对密度
equivalent soil layer method 等值层法(地基沉降)
equivalent specific heat 等效比热
equivalent stress 等效应力
equivalent weight replacement method
　　　　等重量代替法(粗粒土配料)
equivoluminal wave 等体积波
erodibility 冲蚀性

erosion surface 侵蚀面
erratic 漂石
erratic block 漂砾
erratic form 漂积层
erratic soil profile 不规则土层剖面
erratic soil structure 土的不规则结构
erratic subsoil 杂乱土层，不均一土
eruptive rock 喷出岩
escarpment 陡崖
estavel 地下河
estuary deposit 河口沉积
Euler crippling stress 欧拉断裂应力
evaluation of opening stability
　　　　　峒室稳定性评价
evaporation 蒸发
evapo-transpiration 蒸发总量
eventual flood 特大洪水
ever frozen layer 永冻层
ever frozen soil 永冻土
excavated pile 挖孔桩
excavated section 开挖断面
excavated volume 挖方
excavating machinery 挖掘机械
excavation 挖方，开挖
excavation depth 开挖深度
excavation slope 挖方坡度
excavator 挖土机
excess hydrostatic pressure 超静水压力
excess load 超载
excess pore water pressure, u_e
　　　　　超静水压力，超孔隙水压力
excessive settlement 过大沉降
exchangeable ion 可交换离子
expanded clay 膨胀黏土

expanded pearlite 膨胀珍珠岩
expanded polystyrene [EPS] 聚苯乙烯发泡材料
expanded shale 膨胀页岩
expanding auger 扩孔钻
expanding bit 扩孔钻头
expansibility 膨胀性
expansion 膨胀
expansion apparatus 膨胀仪
expansion index 回弹指数(压缩试验)
expansion joint 伸缩缝
expansion shell bolt 膨壳式锚杆
expansive force 膨胀力
expansive pressure 膨胀压力
expansive soil 膨胀土
experimental data 试验数据
experimental parameters 试验参数
experimental soil engineering 试验土工学
experimental study 试验研究
experimental table 试验台
experimentation 试验(方法)
exploded pile 爆扩桩
exploration 勘探
exploration drilling 钻探
exploration investigation 探索性研究
exploration stage 勘探阶段
exploration survey 勘测
exploration trench 探槽
exploration work 勘探工作
exploratory 勘探
exploratory adit 勘探平洞
exploratory bore-hole 勘探钻孔
exploratory line 勘探线
exploratory pit 探坑
exploratory shaft 探井

explosive compaction 爆炸挤密
express pile 大头桩
extension test 拉伸试验
extensometer 伸长计，应变计
external force 外力
extractor force 拔桩力
extra-sensitive clay 超灵敏黏土
extraterrestrial soil 地球外的土
extruder 顶样器，挤压机
extrusive rock 喷出岩

F

fabric 组构
fabric analysis 组构分析
fabric anisotropy 组构各向异性
fabric forms 织物模板
fabric retaining wall 土工布挡土墙
fabric sheet reinforced earth 铺网法
fabric-enclosed sand drain 袋装砂井
face drain 表面排水
face shovel 正向铲
facing 面板
factor of reduction 折减因素(桩工)
factor of safety 安全系数
factor of safety against overturning
　　　抗倾覆安全系数
factor of safety against sliding 抗滑安全系数
fading of settlement 沉降衰减
fail in compression 压缩破坏
fail in shear 剪切破坏
fail in tension 拉伸破坏
failure by heaving 隆起破坏
failure by piping 管涌破坏
failure by sub-surface erosion 地下侵蚀破坏
failure condition 破坏条件
failure criterion 破坏准则
failure envelope 破坏包线
failure hypothesis 破坏假说
failure index 破坏指数
failure load 破坏荷载
failure locus 破坏轨迹
failure mechanism 破坏机理

failure of earth slope 土坡滑塌
failure of rock mass 岩体破坏
failure plane 破坏面
failure strain 破坏应变
failure strength 破坏强度
failure stress 破坏应力
failure surface 破坏面
failure test 破坏试验
failure zone 破坏区
fall 落距(锤)，倒塌(岩，土)
fall-cone test 落锤试验
fall of ground 冒顶
falling cone method 沉锤法(液限试验)
falling head permeability test
　　　　变水头渗透试验
falling head permeameter 变水头渗透仪
fascine mattress 沉排
fascine revetment 沉排护岸
fast compression test 快速压缩试验
fat clay 肥黏土
fathometer 回声测深仪
fatigue 疲劳
fatigue failure 疲劳破坏
fatigue fracture 疲劳断裂
fatigue limit 疲劳界限
fatigue loading 疲劳荷载
fatigue resistance 抗疲劳强度
fatigue shear apparatus 疲劳剪切仪
fatigue strength 疲劳强度
fault 断层
fault block 断层石块
fault gouge 断层泥
fault plane 断层面
fault zone 断层带

feasibility design 可行性设计
feasibility study 可行性研究
federal aviation agency classification [FAAC] 中央航空局土分类法
feedback control 反馈控制
feeler inspection 探针检验
feldspar 长石
feldspathic quartz sandstone 长石石英砂
feldspathic sandstone 长石砂岩
Fellenius method of slices 费伦纽斯条分法
Fellenius solution 费伦纽斯解法(地基承载力)
fen land 沼泽地
fen soil 沼泽土
fender pile 护桩
ferrohydrite 褐铁矿
ferruginous cement 铁质胶结物
ferruginous lateritic soil 含铁红壤
fiberfill 纤维填充物
fiber-reinforced concrete 纤维加强混凝土
fibrous peat 纤维性泥炭
fibrous soil 纤维性土
field bearing test 现场载荷试验
field blasting test 现场爆炸试验
field compaction curve 现场压实曲线
field compaction test 现场碾压试验
field compressometer 螺旋压板载荷试验仪
field control 现场控制
field data 现场数据
field density test 现场密度试验
field exploration 野外勘探
field groundwater velocity 地下水实际流速
field identification 现场鉴定
field identification of soil 土的现场鉴别
field investigation 现场勘察

field laboratory 现场试验室
field loading test 现场载荷试验
field measurements 现场测试
field moisture 天然湿度,天然含水量
field moisture equivalent 现场含水当量
field observation 现场观测
field operation 现场作业
field permeability test 现场渗透试验
field pumping test 现场抽水试验
field recompression curve 现场再压缩曲线
field sounding test 现场触探试验
field test 现场试验室
field vane test 现场十字板试验
field work 野外作业,外业
fill 填方,填土
fill(ed) ground 填土地基
filled pile 灌注桩
filled soil 填土
film water 薄膜水
filter 反滤层
filter blanket 反滤铺盖,滤水垫层
filter cloth 滤布,滤网
filter fabric mat 土工布反滤层,反滤织物层
filter fabric soil retention test
　　　　土工布反滤层试验,滤层织物阻土试验
filter layer 滤层,反滤层
filter mat 滤水毡垫
filter paper drain 滤纸排水
filtering flow 滤流
filtering layer 滤层,反滤层
filter-well 渗水井
filtration 过滤
filtration erosion test 渗透变形试验
final penetration 最终贯入度(桩工)

final set 最终贯入度(桩工)
final settlement 最终沉降(量)
final strength 最终强度
final stress 有效应力
fine aggregate 细骨料，细粒料
fine analysis 细颗粒分析
fine grained soil, fines 细粒土
fine rock fissure 岩石细裂隙
fine sand 细砂
fine silt 细粉土
fine silt bond 细粉粒粘结
fine-graded 细级配的
fine-grained soil 细粒土
fineness modulus 细度模数
fineness of grain 颗粒细度
finger stone 小石块
finished ground level 竣工地面高程
finite difference method 有限差分法
finite element method [FEM] 有限元法
finite slice method 条分法
firm clay 硬黏土
fish 钻孔异物
fissibility 易裂性
fissure 裂隙
fissure aperture 裂隙张开度
fissure water 裂隙水
fissured clay 裂隙黏土，龟裂(的)黏土，裂缝土
fissured rock 裂隙岩石
fitting method 拟合法
fixed ground water 固定地下水
fixed piston sampler 固定活塞式取样器
fixed-ring consolidometer 固定环式固结仪
fixing bolt 固定锚杆
flake-shaped particle 片状颗粒

flaky constituents 片状成分
flame retardant fibers 防燃土工布，阻燃织物
flat dilatometer 扁式膨胀仪
flat jack 扁千斤顶
flat jack technique 扁千斤顶法
flatness ratio 扁平度，扁平比，平滑比
flexible conduit 柔性管道
flexible foundation 柔性基础
flexible load 柔性荷载
flexible membrane liner 柔性薄膜衬垫
flexible pavement 柔性路面
flexible wall 柔性墙
flexural rigidity 抗挠刚度
flexural stress 挠曲应力
flint 燧石
flint clay 硬质黏土
floating caisson 浮运沉箱
floating crane 浮吊
floating foundation 浮筏基础
floating pile 摩擦桩
floating pile driver 水上打桩机
floating pile foundation 摩擦桩基础
float-ring consolidometer 浮环式固结仪
floc 絮状物
flocculated clay buttress bond 絮状支托粘结
flocculation 絮凝作用
flocculent soil 絮凝土
flocculent structure 絮凝结构
flood deposit 洪积物
flood plain deposit 泛滥平原沉积
flow channel 流槽
flow curve 流动曲线（液限试验）
flow failure 流动破坏
flow gauge 流量计

flow index 流动指数
flow lane 流槽
flow layer 流动层
flow line 流线
flow net 流网
flow path 流径
flow phenomenon 塑流现象
flow rock 流岩，流砂
flow rule 流动法则
flow slide 泥流型滑坡，塑流型滑坡
flow structure 流状构造
flow value 流值
flowing artesian 自流承压水
flowing avalanche 流动(滑动)崩坝
flowing ground 流动性地基(流砂)
flowing sand 流砂
fluctuation belt of water table 地下水位变动带
fluidification 液化
fluidised bed 流砂地层
flutter 振动
fluvial deposit 河流沉积
fluvial erosion 河流冲蚀
fluvial soil 河流冲积土
fluvial terrace 河成阶地
fluvio-glacial deposit 冰水沉积
fluvio-marine deposit 河海沉积
flyash 粉煤灰
flyash stabilized soil 粉煤灰加固土
foil sampler 衬片取样器
fold 褶皱
folded rock 褶皱岩石
foliation 叶理，剥理
follower 送桩
footing 基础

footing analogy   等效基础模拟
footing beam   基础梁
forced vibration   强迫振动
forcing hammer foundation   锻锤基础
forcing pump   压力泵
foreshock   前震
formation   地层，建造(地质)
formation level   岩层面，路基面
formation pressure   地层压力
former   打桩套管
fossil landslide   古滑坡
fossil soil   古土壤，化石土
foundation   基础，地基
foundation analysis   基础分析
foundation anchoring   基础锚固
foundation base   基底
foundation beam   基础梁
foundation bed   基床，基础垫层
foundation bolt   地脚螺栓
foundation by means of injecting cement
    基础灌浆加固
foundation by pit sinking   挖坑沉基
foundation characteristics   基础特征
foundation construction   基础施工
foundation course   基层
foundation deformation   基础变形
foundation design   地基设计，基础设计
foundation design code for building
    建筑地基基础设计规范
foundation dewatering   基础降水
foundation displacement   基础位移
foundation ditch   基础沟，基(础)坑，基槽
foundation drain hole   基础排水孔
foundation embedment   基础埋置深度

foundation engineering 基础工程，基础工程学
foundation excavation 地基开挖
foundation exploration 地基勘探
foundation failure 基础破坏
foundation footing 基脚
foundation grouting 地基灌浆
foundation improvement 基础加固
foundation investigation 地基勘察
foundation isolation 基础隔振
foundation layout plan 基础布置平面图
foundation leakage 地基渗漏
foundation level 基底标高
foundation line 基础线，开挖线
foundation load 基础荷载
foundation made of materials 三合土基础
foundation mat 基础底板，基褥垫，基垫层
foundation modulus 地基模量，地基反力系数
foundation of masonry 圬土基础
foundation pad 基础垫层
foundation pile 基础桩，基桩
foundation pit 基坑
foundation pit drainage 基坑排水
foundation platform 基础承台
foundation practice 基础施工技术
foundation pressure 基础底面压力，基底压力
foundation resilience test 地基回弹测试
foundation settlement 地基沉降，基础沉陷
foundation slab 基础底板，承台
foundation soil 地基土
foundation stabilization 地基加固
foundation stone 基石
foundation strata 地基土层
foundation structure 基础结构
foundation treatment 地基处理，基础处理

foundation trench 基槽
foundation under water 水下基础
foundation underpinning 基础换托
foundation vibration 基础振动
foundation wall 基础墙,基墙
foundation work 地基工程,基础工程
foundations 地基
four bladed vane 十字板仪
four-sheet mineral 四片层矿物
fraction 粒组
fraction of partial size 粒径组
fraction toughness of rock 岩石断裂韧度
fracture 断裂,破裂
fracture grouting 劈裂灌浆
fracture mechanics 断裂力学
fracture plane 断裂面
fracture zone 断裂破碎带,破裂带
fragment 碎屑
frame foundation 框架式基础
framework silicate structure 架状硅酸盐结构
Franki pile 法兰基灌注桩
free face 临空面
free period 自由周期
free stone 毛石
free swell test 自由膨胀率试验
free swelling ratio 自由膨胀率
free vibration 自由振动
free vibration column test 自振柱试验
free water 自由水
free water elevation 地下水位,自由水位
free-end triaxial test 自由端三轴试验
freeze 歇后增长(桩的承载力)
freeze-thaw resistance 抗冻融强度
freeze-thaw test 冻融试验

freezing force 冻结力
freezing index 冻结指数
freezing isoline 等冻结线
freezing method 冻结法
freezing point 冰点
freezing process 冻结法
freezing sampler 冻结取样器
freezing zone 冻结带
french drain 盲沟
frequency analysis 频谱分析
frequency at resonance 共振频率
frequency-amplitude curve 频率-振幅曲线
frequency-response curve 频率特性曲线
fresh rock 新鲜岩石
fresh water 淡水
freshet 山洪
friable rock 松散岩石
friable soil 松散土，酥性土
friction angle 摩擦角
friction circle method 摩擦圆法
friction coefficient 摩擦系数
friction damping 摩擦阻尼
friction lag 摩擦滞后
friction loss 摩阻损失
friction pile 摩擦桩
frictional resistance 摩擦阻力，摩阻力
frictional resistance coefficient 摩擦阻力系数
frictional soil 无黏性土
friction-resistance ratio 摩阻比
frog-rammer 蛙式打夯机
frontiers 拓宽(地基处理)
frost action 冻结作用
frost boil 翻浆，冻融
frost depth 冻结深度

frost effect 冻结作用
frost front 冻结前缘
frost heave 冻胀
frost heaving pressure 冻胀力
frost line 冰冻线
frost penetration 冻结深度
frost pressure 冻胀压力
frost resistance 抗冻力
frost strength 冻结强度
frost susceptibility 易冻性
frost zone 冻结区
frosted glass 毛玻璃
frost-heave capacity 冻胀量
frost-heaving pressure 冻胀力
frost-heaving soil 冻胀土
frost-proof depth 防冻深度
frost-susceptible soil 易冻土
frozen heave test apparatus 冻胀仪
frozen soil 冻土
frozen-ground phenomenon 地冻现象
frozen-heave force 冻胀力
fry-dry method 炒干法
full face method 全断面法(隧道)
full pore-pressure ratio 全孔隙压力比
full scale test 足尺试验
full size 实际尺寸
fullness coefficient 充盈系数
fully softened strength of clay 黏土完全软化强度
funnel 漏斗
fuse 引线

G

gabbro 辉长石
gallet 石屑
gamma-ray density gauge γ射线密度仪
gamma-ray log γ射线测井
gang of wells 井群(组),组合井
gap gradation 不连续级配
gap-graded soil 不连续级配土
gap grading 间断级配
gapping 缩颈(桩工)
gapping fissure 张开裂度
garbage dump 垃圾填土
gas content 含气量
gaseous phase 气相
gauge of bit 钻头规格
Gauss normal distribution curve
　　高斯正态分布曲线
gel 凝胶
general shear failure 整体剪切破坏
generalized procedure of slices 广义条分法
general-shear failure 整体剪切破坏
genetic soil 原生土
geocomposite 土工复合材料
geodrain(prefabricated strip drain)
　　塑料排水带法
geofabriform 土工模袋
geogrid 土工格栅,网格形土工织物
geohydrology 地质水文学
geoisotherms 等地温线
geologic environment 地质环境
geologic environment element 地质环境要素

geologic features　地质特征
geologic investigation　地质勘察
geologic structure　地质构造
geological age　地质年代
geological chronology　地质年代学
geological column　地质柱状图
geological condition　地质条件
geological defect　地质缺陷
geological discontinuity　地质不连续面
geological exploration　地质勘探
geological history　地质历史
geological log　地质记录，地质柱状图
geological map　地质图
geological observation point　地质点
geological origin　地质成因
geological period　地质年代
geological profile　地质剖面图
geological prospecting　地质勘探
geological report　地质报告
geological section　地质剖面图
geological structure　地质构造
geological succession　地质演变，地层系统
geological suitability of site　场地地质适宜性
geological survey　地质调查
geologist　地质学家，地质人员
geologists' pick　地质锤
geology　地质学
geomat　土工垫
geomechanical model test　地质力学模型试验
geomechanics　地质力学，岩石力学
geomembrane　土工膜
geometric head　几何学水头
geometrical approximation　几何近似
geometrical damping　几何阻尼

geometrical isotropy 几何各向同性
geometrical mean 几何平均
geometrical similarity method 几何相似法
geomorphology 地貌，地貌学
geonets 土工网
geophone 拾振器
geophysical exploration 地球物理勘探
geophysical log 地球物理测井
geophysical prospecting 物探
geophysics 地球物理学
geopolymer 土工聚合物
geostatic pressure 地压力
geostatic stress 自重应力，地应力
geostress 地压力
geosyncline 地槽
geosynthetics 土工合成材料
geotechnical centrifugal model test
　　　　土工离心模型试验
geotechnical engineering 岩土工程(学)
geotechnical engineering and underground engineering
　　　　岩土与地下工程
geotechnical engineering test 岩土工程测试
geotechnical exploration 岩土工程勘探
geotechnical fabrics(geofabric) 土工布，土工
　　　　　　　　　　　　　　织物
geotechnical geology 大地构造学
geotechnical grouting 岩土灌浆
geotechnical investigation
　　　　岩土工程勘察
geotechnical model test 土工模型试验
geotechnical processes 岩土工程方法
geotechnician 岩土工程师
geotechnics 大地构造学
geotechnique 岩土工程，岩土工程技术

geotechnology 岩土工程学
geotextiles 土工织物
Gibbson soil 吉布森地基模型
glacial breccia 冰川角砾
glacial clay 冰碛黏土
glacial deposit 冰川沉积
glacial erosion 冰蚀
glacial loess 冰川黄土
glacial outwash 冰水沉积
glacial rock 冰成岩
glacial till 冰碛土
glacial-fluvial soils 冰水沉积土
glacier 冰川
glaciology 冰川学
gneiss 片麻石
Goble pile-driving analyser 戈布尔打桩分析器
gouge 断层泥
Gow caisson 多级套筒式沉井
grab tensile strength 抗抓拉强度
grab tensile test 抓拉试验
grabbing excavator 抓斗挖土机
gradated filter 级配反滤层
gradated soil 级配土
gradation 级配，修坡
gradation test 粒径分析
grader 平土机
gradient 梯度
gradient ratio test 梯度比试验
grading 级配，配级，修坡
grading analysis 粒径分析
grading collapse settlement 分级湿陷量
grading curve 粒径分布曲线
grading factor 级配系数
grading fraction 粒径分级

graduated cylinder 量筒
grain 颗粒，土粒
grain boundary 粒径界限
grain breakage 颗粒破碎率
grain diameter 粒径分级
grain pressure 粒间压力
grain shape 颗粒形状
grain size 粒径
grain size accumulation curve 粒径累积曲线
grain size composition 粒径组成
grain size distribution curve 粒径分布曲线
grain size measurement 颗粒度分析
grain skeleton 颗粒骨架
grainage 粒度
graining sand 细砂
grainness 粒度
grain-size analysis(granulometry) 颗粒分析，粒径分析
grain-size analysis curve 颗粒分析曲线
grain-size curve 粒径曲线
grain-size distribution 粒径分布
grain-size distribution curve 粒径分布曲线
grain-size frequency curve 粒径频率曲线
granite 花岗石
granular soil 粒状土
granularity 粒度
granularmetric analysis 粒径分析
granulometric composition 粒径组成
graphical analysis 图解分析
graphical representation 图示法
gravel 砾石
gravel cobble 卵石
gravel drain 砾石排水沟
gravel foundation 砾石基础

gravel pack 砾石填充层
gravel pile 碎石桩
gravel-filled drain trench 砾石排水沟
gravelly sand 砾砂
gravelly soil 砾质土
gravelly soil foundation 碎石土地基
gravel-sand cushion 砂砾垫层
gravimeter 重力仪
gravitational exploration 重力勘探
gravitational potential 重力势
gravitational water 重力水，自由水
gravity fault 重力断层
gravity hammer 重力锤
gravity retaining wall 重力式挡土墙
gravity stress 自重应力
green sand 新采砂
grid-mat foundation 格筏基础
Griffith's strength criterion 格里菲斯强度准则
grillage foundation 格排基础
grillage raft 格排筏式基础
griotte 大理石
grit 粗砂，砂岩
grit stone 粗砂岩
gritty soil 粗砂砾质土
groin 丁坝
gross analysis 全量分析
gross loading intensity 基底压力，总荷载强度
ground acceleration 地面加速度
ground anchor 地层锚杆
ground arch 塌落拱
ground base 地基
ground beam 地基梁，地梁
ground bearing pressure 地基承载力

ground compaction 地基压缩
ground condition 地基条件
ground consolidation 地基固结
ground engineering 地基工程，地面工程
ground fracturing 地裂
ground freezing 地基冻结法
ground improvement 地基加固
ground level 地面标高
ground loss 地面下陷
ground map 地形图
ground motion 地面运动
ground movement 地动
ground pressure 地层压力
ground settlement 地面沉降
ground slope factor 地面倾斜因素（地基承载力）
ground sluice 地沟
ground stabilization 地基加固
ground stress 地应力
ground subsidence 地面下沉
ground surface 地表，地面
ground treatment 地基处理
ground vegetation 地表植被
ground work 基础，地基
groundwater 地下水，潜水，自由水
groundwater cascade 地下水瀑布
groundwater control 地下水控制
groundwater discharge 地下水排泄，地下水流量
groundwater divide 地下水分水岭
groundwater dynamics 地下水动力学
groundwater elevation 地下水位
groundwater evaporation 地下水蒸发
groundwater hardness 地下水硬度
groundwater increment 地下水补给量
groundwater isopiestic line 地下水等位线

groundwater level(table) 地下水位
groundwater lowering 降低地下水
groundwater monitoring 地下水监测
groundwater pollution 地下水污染
groundwater recharge 地下水补给量
groundwater regime 地下水动态
groundwater ridge 地下水分水线
groundwater steady flow 地下水稳定流
groundwater storage 地下水贮存量
groundwater tracer test 地下水跟踪试验
groundwater unsteady flow 地下水不稳定流
group action 桩群作用
group efficiency 群桩效率
group index 分组指数(路工)
group piles 群桩
grout 浆液
grout absorption 吸浆量,吃浆量
grout acceptance 吸浆量,吃浆量
grout agitator 浆液搅拌机
grout box 灌浆盒
grout cap 灌浆盖板,灌浆帽
grout cohesion 灌浆凝聚力
grout consumption 耗浆量
grout curtain 灌浆帷幕
grout delivery 供浆量
grout discharge valve 浆液排放阀
grout distribution line 配浆管路
grout effectiveness 灌浆效果
grout formulation 浆液配方
grout gallery 灌浆廊道
grout hole 灌浆孔
grout injection pipe 灌浆管
grout mix 浆液配比
grout mix design 浆液配合比设计

grout outlet 出浆口
grout pipe 灌浆管
grout take 耗浆量
groutability 可灌性
groutability ratio 可灌比
grouted anchor 灌浆锚杆
grouted area 灌浆区
grouted base pile 基底灌浆桩
grouted bolt 灌浆锚杆
grouted cut-off wall 灌浆截水墙
grouted riprap 灌浆乱石
grouted-aggregate concrete 骨料灌浆混凝土
grout-filled fabric mat 灌浆土工布护排
grouting 灌浆
grouting machine 灌浆机
grouting material 灌浆材料
grouting operation 灌浆作业,灌浆施工
grouting parameter 灌浆参数
grouting pressure 灌浆压力
grouting project 灌浆工程
grouting pump 灌浆泵
grouting radius 灌浆半径
grouting result 灌浆成果
grouting stage 灌浆段
grouting test 灌浆试验
grubbing 清理场地
guard wall 护墙
guide adit 导洞
guide fossil 标准化石,主导化石
guide pile 导桩
guide runner 导板
guide trench 导沟
guide tube 导管
guide wall 导墙

gulch 冲沟，干谷，峡谷
gulley 集水沟，排水沟
gully 冲沟
gully erosion 沟蚀
gumbo 肥黏土(美国)
gunite 喷浆
gusher type well 自流井(自流钻井)
Guttman process 谷特曼法(化学加固)
gypsum 石膏
gyttja 腐殖泥

H

hair crack 毛细裂缝
half-space 半空间
halloysite 埃洛石，多水高岭石
halomorphic soil 盐碱土
hammer cushion 桩锤垫
hand auger 手钻
hand boring 手工钻探
hand rammer 人力夯
hand-dug 人工开挖
hand-dug well 人力挖井
handling stress 起吊应力(桩工)
hand-operated auger 手动螺旋钻
hanging curtain 悬挂式帷幕
hard clay 硬黏土
hard water 硬水
harden grout film 水泥结石
harden modulus 硬化模量
hardening 硬化
hardening parameter 硬化参数
hardening rule 硬化规则

hardness 硬度
hardness degree of rock 岩石坚硬程度
hardpan 硬土层
harmonic vibration 简谐振动
Harvard miniature compaction test
    哈佛小型碾实试验
hauling scraper 铲运机
head loss 水头损失
header pipe 总管
heading 导洞
heat bonding 热粘合
heat capacity 热容量
heat convection 热对流
heat radiation 热辐射
heat transfer 热传导
heat treatment 热处理
heave 隆胀
heave compensator 浮升补偿器
heave force 隆胀力
heave stake 隆起标桩
heaving of the bottom 基坑底隆胀
heavy foundation 重型基础
heavy grade 陡坡
heavy minerals 重矿物
heavy soil 重黏土
heavy spar 重晶石
heavy tamping 重锤夯实
heavy textured soil 重粘结土，重黏土质
height of drop 落距
held water 结合水
helical auger 螺旋钻
helmet 桩帽
hematite 赤铁矿
heterogeneity 非均质性

heterogeneous soil 非均质土
heterogeneous stratum 非均质地层
high air entry piezometer tip 高进气孔隙水压力测头
high liquid limit soil 高液限土
high pressure grouting 高压灌浆
highly compressed clay 高压缩黏土
highly plastic soil 高塑性土
high-order elastic modulus 高阶弹性模量
high-rise pile cap 高桩承台
highway embankment 公路路堤
Hiley formula 希利公式(桩工)
hillside 小山坡
histogram 直方图,柱状图
Histosol 有机土
hoggin 夹砂砾石
holard 土壤水
hole diameter measurement 井径测量
hole sealing 封孔
hole spacing 孔距
hole wall 孔壁
hollow cylinder test 空心圆柱体试验
Holocene Epoch 全新世
holographic interferometry 全息干涉测量
home 停打阻力(桩工)
homogeneous soil 均质土
homogeneous stratum 均质层
honeycomb structure 蜂窝结构
Hooka's law 虎克定律
Hookean body 虎克体
Hookean model 虎克模型
hoop stress 环向应力
horizontal bore hole drain 水平钻孔排水
horizontal filter well 水平滤井

horizontal heave force 水平冻胀力
horizontal movement gauge 水平位移计
horizontal plate gauge 板式水平位移计
hornblende 角闪石
host rock 主体岩石
hovering pressure 滑动压力
H-pile H桩
human erosion 人为侵蚀
humus soil 腐殖土
hundred percent pore pressure ratio 100%孔压比
Hvorslev parameter 沃斯列夫参数
Hvorslev soil model 沃斯列夫模型
Hvorslev surface 沃斯列夫面
hydatogen sediment 水成沉积
hydatogenous rock 水成岩
hydrated bentonite 水化膨润土
hydrated ion 氢氧离子
hydrated lime 熟石灰
hydrated mineral 水化矿物，含水矿物
hydration 水合作用，水化作用
hydration heat 水化热
hydration water 吸附水
hydrauger hole 水冲钻孔
hydraulic conductivity 透水性，水力传导系数
hydraulic fill 水力冲填，吹填
hydraulic fracture 水力劈裂，水力破坏
hydraulic fracturing technique 水力劈裂法
hydraulic gradient 水力梯度
hydraulic jack 液压千斤顶
hydraulic levelling device 流体测沉计
hydraulic load cell 水力压力盒
hydraulic overflow settlement cell 溢水式沉降计
hydraulic percussion method 水力冲击钻探法

hydraulic permeability 水力透水性
hydraulic slope 水力坡度
hydraulic tunnel 水工隧道
hydro-blasting 加水爆破法（黄土），水爆法
hydrogeological drilling 水文地质钻探
hydrogeological investigation 水文地质勘察
hydrogeology 水文地质学
hydroisohypse 等水深浅
hydrometer analysis 比重计分析
hydromica 水云母
hydromophic soil 水成土
hydrophilic colloid 亲水胶体
hydrophobic colloid 憎水胶体
hydroscopic moisture 吸湿性
hydroscopic water 吸附水
hydrosiderite 褐铁矿
hydrostatic distribution 静水压力分布
hydrostatic excess pressure 超静水压力
hydrostatic head 水头
hydrostatic line 静水压力线
hydrostatic pressure 静水压力
hydrostatic profile gauge 静水位移监测计
hydrostatic state of stress 静水应力状态
hydrostatic uplift 静水上托力
hygrometer 湿度计
hygroscopic capacity 吸湿容量
hygroscopic coefficient 吸湿系数
hygroscopic water content 吸湿含水量
hyperbolic model 双曲线模型
hyperelastic law 超弹性定律
hypocenter 震源
hypoelastic law 准弹性定律
hypoelastic model of soil 土的次弹性模型
hypogene rock 深成岩

hysteresis compaction 滞回压实
hysteresis loop 滞回环
hysteresis modulus 滞回模量
hysteretic elastic model of soil 土的滞后弹性模型

I

ice content 含冰率
ice lenses 冰透镜体
ice pressure 冰压力
ice-laid deposits 冰川沉积
ideal elastoplastic model 理想弹塑性模型
ideal grading curve 理想颗粒级配曲线
ideal liquid 理想液体
Idel sonde 艾德测头
identification of soils 土的识别
igneous rock 火成岩，岩浆岩
ignition loss 烧失量
illite 伊利石
illuvial soil 淀积土壤
immature residual soil 新残积土
immediate compression 瞬时压缩
immediate settlement 初始沉降，瞬时沉降
immersed tube method 沉管法
immersed tunnel 沉埋式隧道
immersion test 浸渍试验
impact block 冲击块
impact center 冲击中心
impact compaction 冲击压实
impact crusher 冲击式破碎机
impact crushing value test 冲击压碎值试验
impact energy 冲击能

impact factor　冲击系数
impact hammer　冲锤
impact load　冲击荷载
impact load of ship or raft　船只或排筏冲击力
impact strength index　冲击强度指数
impact strength test　冲击强度试验
impaction roller　夯击碾，夯实碾
impedance ratio　阻抗比
impeded drainage　排水不良
imperfect pile　缺陷桩
impermeability　抗渗性
impermeability test　抗渗试验
impermeable barrier　不透水层
impermeable layer　不透水层，隔水层
impervious　不透水的
impervious barrier　防渗体
impervious blanket　防渗铺盖
impervious boundary　不透水边界
impervious core　不透水心墙，防渗心墙
impervious curtain　防渗帷幕，不透水帷幕
impervious foundation　不透水地基
impervious layer　不透水层，隔水层
impervious liner　不透水衬层
impervious stratum　不透水地层
improvement of liquefiable soil　液化地基加固
impulse load　脉冲荷载
impulse seismic device　激震装置
incident wave　入射波
incipient failure　初始破坏
inclination correction factors　荷载倾斜因数
inclination method　倾角计算法
inclination observation　倾斜观测
inclination of ground　地面坡度
inclination of load　荷载倾角

inclined fault 倾斜断层
inclined load 倾斜荷载
inclined pile 斜桩
inclined shaft 斜井
inclinometer 测斜仪
inclusion 内含物
incompetent rock 弱胶结岩层
incomplete consolidation 不完全固结
incompressibility 不可压缩性
increment of plastic strain 塑性应变增量
increment(al) load method 荷载增量法
incremental initial strain method 增量初应变法
incremental initial stress method 增量初应力法
incremental launching method 顶推法施工
incremental stiffness 增量刚度
incrustation 结壳
index of strength 强度指标
index property test 指标特性试验
indirect drainage 间接排水
indirect tensile test for rocks 岩石间接拉伸试验
individual footing 单独基础,独立基础
individual stone column 单根碎石桩
indoor test of geotechnique 岩土室内试验
induce stress 诱发应力
induced earthquake 诱发地震
industrial contaminant 工业污染物
industrial waste 工业废料
inelastic buckling 非弹性屈曲
inelastic collision 非弹性碰撞
inelastic deformation 非弹性变形
inertia 惯性,惯量

inertia force 惯性力
inertia principal axis of area 截面主惯性轴
infiltration 渗透，渗滤
infiltration capacity 渗入量
infiltration path 渗透路径
infiltration point 入渗点
infiltration rate 渗透率
infiltration test 渗水试验
infiltration velocity 渗透速度
infiltration well 渗水井
infinite element method 无限元法
infinite slope 无限边坡
inflation pressure 膨胀压力
inflection knee(point) 拐点
inflection point method 拐点法
inflexibility 刚性
influence chart 感应图
influence coefficient 感应系数
influence factor for axial force 轴向力影响系数
influence factor for eccentricity 偏心影响系数
influence value 感应值
influent seepage 渗漏
infrared analysis 红外光分析
infrared detection 红外探测
ingress of water 水的进入
inhomogeneity 非均质性
inhomogeneous soil 非均质土
initial collapse pressure 湿陷起始压力
initial compression 初始压缩
initial condition 初始条件
initial consolidation 初始固结
initial consolidation pressure 初始固结压力
initial eccentricity 初偏心

initial imperfection 初始缺陷
initial liquefaction 初始液化
initial looseness coefficient 初始可松性系数
initial moisture content 初始含水量
initial overburden pressure 初始超载压力
initial parameter method 初参数法
initial placement condition 初始建筑条件
initial plastic flow 初始塑性流动
initial pore water pressure 初始孔隙水压力
initial rock stress field 原岩应力场
initial setting 初凝
initial settlement 瞬时沉降
initial shear modulus 初始剪切模量
initial shear resistance 初始剪切阻力
initial shear stress ratio 初始剪应力比
initial strength 初始强度
initial stress 初始应力
initial tangent modulus 初始切线模量
initial value problem 初值问题
initial void ratio 初始孔隙比
injectable soil 可灌地基土
injecting grout 灰泥注浆
injection 灌入,喷射
injection well 注水井,回灌井
injection flow 灌注流量
injection grout 压力灌浆,注入浆
injection material 灌浆材料
injection pipe 灌浆管,喷射管
injection pressure 灌浆压力
injection process 注浆法,灌浆法
injection pump 射流泵,注浆泵
ink bottle effect 瓶颈效应
inlier 内围层(岩石)
inner drainage 内排水

inner force 内力
inorganic soil 无机土
in-place density 原位密度
in-place permeability 地层渗透率
input well 注水井,回灌井
insequent landslide 切层滑坡
inside clearance ratio 内间隙比(取样器)
in-situ CBR test 原位加州承载比试验
in-situ concrete pile 现浇混凝土桩
in-situ direct shear test 原位直剪试验
in-situ monitoring 现场监测
in-situ pile 就地灌注桩
in-situ soil 原状土
in-situ soil test 原位土工试验
in-situ soil testing 土的原位试验
in-situ stabilized column 原位加固土柱
in-situ strength 现场强度
in-situ stress 原位应力
in-situ test(s) 原位试验,原位测试
in-situ test of geotechnique 岩土现场试验
in-situ thrust test 原位推力试验
inspection borehole 检查孔
inspection pit 探坑
inspection well 检查井
instability 不稳定性
instability index 不稳定指数
instability of first kind 第一类失稳
instability of second kind 第二类失稳
instantaneous elastic strain 瞬时弹性应变
instantaneous load 瞬时荷载
instantaneous pore pressure 瞬时孔隙压力
instantaneous rupture 瞬时破坏
instantaneous stability of slope 边坡瞬时稳定
instrument gallery 观测廊道,仪表廊

instrumental error 仪器误差
instrumented pile 装有量测元件的桩
intact clay 原状黏土
intact ground 未扰动地基
intact rock 完整岩石
intact specimen 原状试样
intactness index of rock mass 岩体完整性指数，岩体速度指数
integral coring method 整体取芯法
integral rigidity 整体刚度
integrity testing 完整性检测(桩基)
intelligent grouting parameter recorder 智能灌浆参数记录仪
intensity adjustment 烈度调整
intensity assessment 烈度评定
intensity curve 强度曲线
intensity determination 烈度测定
intensity distribution 烈度分布
interaction effect 相互作用
interactive coefficient 相互影响系数
interatomic bonding force 原子键力
interbedded strata 互层
interbedding 互层
intercalation 夹层
intercepting well 截流井
interceptor drain 截水沟
interface 分界面，接触面
interface speed 界面速度
interface strength 接触面强度
interfacial bond strength 界面粘结强度
interfacial force 界面力
interfacial friction 层间摩擦
interfacial shear strength 界面剪切强度
interfacial tension 界面张力

interformational sliding 层间滑动
intergelisol 层间冻土
intergranular cement 粒间胶结
intergranular permeability 粒间透水性
intergranular pressure 粒间压力
interlaminar shear strength 层间剪切强度
interlayer 隔层，夹层
interlayer bonding 层间联结
interlayer contact 层间接触
interlayer continuity 层间连续性
interlayer spacing 层间距
interlayer water 层间水
interlayer-gliding fault 层间滑动断裂
interlocking action 咬合作用
intermediate grain 中级颗粒
intermediate principle plane 中主平面
intermediate principle strain 中主应变
intermediate principle stress 中主应力
intermediate rock 中性岩
intermediate water 层间水
intermittent load 间歇荷载
intermolecular bonding force 分子键力
internal cohesion 内黏聚力
internal crack 内部裂缝
internal damping 内阻尼
internal diameter 内径
internal force 内力
internal force redistribution 内力重分布
internal friction 内摩擦
internal friction angle 内摩擦角
internal instrument installation 仪器埋设
internal pressure 内压
internal scour 内部冲刷，潜蚀
internal support 内支护

interparticle attractive force 粒间引力
interparticle bond 粒间键
interparticle force 粒间作用力
interparticle friction 粒间摩擦
interparticle repulsion 粒间斥力
interpermafrost water 冻结层间水
interpretative log 钻探解释剖面
intersection angle 交角
intersection line 交线
intersection plane 交面
intersection point 交点
intersheet bonding 片间粘结
interstice 间隙
interstices of soil 土壤空隙,土壤裂缝
interstitial pressure 孔隙压力,间隙压力
interstitial water 间隙水
interstrated water 层间水
interstratified bed 间层,夹层
intrinsic permeability coefficient 内在渗透系数
intrinsic property 固有特性
intrinsic shear strength 固有抗剪强度
intrinsic shear strength curve 固有抗剪强度包线
intrusive rock 侵入岩
inundation 浸水,淹没
invariant 不变量
invariants of stress tensor 应力张量不变量
inverse analysis 反演分析
inverse analysis of parameters 参数反分析
inverted arch foundation 倒拱基础
inverted beam method 倒梁法
inverted capacity of well 井的吸收容量
inverted filter 反滤层
inverted well 回灌井,吸水井
investigation 勘察

investigation during construction 施工勘察
investigation stage 勘察阶段
inwash 冰川边缘沉积,岸边淤积
ion concentration 离子浓度
ion diffusion 离子扩散
ion exchange 离子交换
ion exchange capacity 离子交换量
ion flow 离子流
ion hydration 离子水化作用
ion substitution 离子替代
ionic bonding 离子键
ionization layer 电离层
ionization potential 电离势
ionized layer 电离层
ironing 压平(强夯)
irreducible(water)saturation 残余(水)饱和度
irregular bedding 不规则层理
irregular curve 不规则曲线
irregular deformation 不规则变形
irregular load 不规则荷载
irregular particle shape 不规则颗粒形状
irregular specimen for rocks 不规则岩石试件
irruptive rock 侵入岩
iskymeter 现场剪切触探仪
isobar 等压线
isochromatic photograph 等色线照片
isochrone 等时孔压线
isoclinic 等倾线
isogeotherm 等地温线
isohume 等湿度线
isohyetal map 降雨量分布图
iso-intensity curve 等强度曲线
isolated foundation 独立基础
isolated interstice 孤立裂隙(岩石)

isolation 隔振
isolation effectiveness 隔振效果
isolation trench 隔振沟
isoline 等值线
isoline method 等值线法
isomorphous substitution 同晶置换
isopachyte 等厚线
isopiestic line of aquifer 含水层等压线
isosmotic pressure line 等渗压线
isostatic 主应力迹线
isostatic anomaly 均衡异常(地壳)
isostatic curve 等压线
isotherm 等温线
isotope 同位素
isotropic consolidation 各向等压固结
isotropic deposit 各向同性沉积物
isotropic hardening 各向同性硬化
isotropic material 各向同性材料
isotropic semi-infinite solid 各向同性半无限体
isotropic soil 各向同性土，均质土
isotropic stress 各向等应力
isotropy 各向同性

J

jack 千斤顶
jack bit 钻头
jack calibration 千斤顶标定
jack losses 千斤顶力损失
jacked pile 压入桩
jacking beam 顶梁，反力梁
jacking force 千斤顶举升力，顶推力
jacking method 顶升法

jacking of foundation 基础抬升
jacking stress 张拉应力
jack-up rig 自升式钻架
Janbu method of slope stability analysis
  詹布法(边坡稳定)
jet boring machine 喷射式钻孔机
jet column 旋喷柱体
jet grouting 喷射灌浆,旋喷灌浆
jet grouting battery 旋喷管
jet grouting column 旋喷桩柱
jet grouting method 高压喷射注浆法
jet grouting stem 喷射灌浆钻杆
jet mixer 喷射管
jet nozzle 喷嘴
jet pile 射水成桩
jet pump 喷射泵
jet stem 喷射杆
jetted pile 射水沉桩,水冲桩
jetting drilling 水力钻探
jetting height 喷射高度
jetting of pile 桩的冲孔
jetting piling 射水打桩法
jetting process 水冲法
jigging platform 振动台
JM type anchor device JM型锚具
JM type anchorage JM型锚具
job layout 工程放线
job site 工地
joint 节理
joint construction 接头施工(地下连续墙)
joint frequency 节理频率
joint gauge(meter) 测缝计
joint grouting 接缝灌浆
joint opening 缝隙

joint plane 节理面
joint roughness coefficient 裂隙糙度系数
joint seam 接缝止水，灌缝
joint set 节理组
joint with hinge 铰接
jointed rock 节理岩体
Joosten process 乔斯登硅化加固
Jurassic period 侏罗纪
juvenile water 原生水，岩浆水

K

K_0-consolidation K_0 固结
K coefficient method K 法（横向受荷桩分析）
K_0 consolidated drained compression test
　　　K_0 固结排水试验
K_0 consolidated undrained compression test
　　　K_0 固结不排水试验
kaiser effect 凯塞效应
kaolin 高岭土
kaolinite 高岭石
karst 喀斯特，岩溶
karst aquifer 喀斯特含水层，岩溶含水层
karst cave 岩溶洞
karst channels 喀斯特溶槽
karst collapse 喀斯特塌陷，岩溶塌陷
karst feature 岩溶地形，岩溶特征
karst funnel 岩溶漏斗
karst land feature 喀斯特地貌
karst pit 岩溶井，喀斯特井
karst topography 岩溶地形
karst water 岩溶水
karstenite 硬石膏

karstic 喀斯特的,岩溶的,石灰岩溶洞
karstic formation 岩溶地层,喀斯特地层
karstic hydrology 岩溶水文学,喀斯特水文学
katamorphism 风化变质
kation 阳离子
kelly bar 方钻杆
Kelvin model 开尔文模型
kentledge 压重
Kepes model 开派斯模型
keraphyllkite 角闪石
kern stone 粗粒砂岩
key bed 标准层
key pile 主桩
key trench 坝机截水墙槽
key wall 齿墙
key well 基准钻孔
K-G model K-G 模型
kicking piece 垫木
kiln dust 窑灰
kinematic hardening 运动硬化,随动硬化
kinematic viscosity 动力黏滞性
kinetic friction 动摩擦
kinetic stability 动力稳定,动力平衡
kneading 搓揉
kneading compaction 搓揉压实
knee 拐点(曲线)
knee brace 斜撑
knife edge 刀口
knitted fabric 针织物
knitted geotextile 编织土工布
knuckle 坡度突变
Kronecker delta 克罗内克符号 δ
Kullenberg sampler 海底取样器

L

laboratory soil mixer 试验室土样拌合器
laboratory soil tests 室内土工试验
laboratory test 室内试验
laboratory vane test 室内十字板试验
lacustrine deposits 湖相沉积
ladder trencher 多斗挖槽机，链斗挖槽机
ladder wall 梯式加筋锚定墙
Lade yield criterion 拉特屈服准则
lag pile 套桩
lagging 挡土板
lagging of pile 桩箍
lagoonal deposit 潟湖沉积
lahar 火山泥流
laitance 浮浆层，浮浆皮
lake marl 湖成泥灰岩
lake mud 湖泥
lambda method λ法(桩工)
lamella structure 薄层构造
lamina 层状体
laminar flow 层流
laminated soil 层状土
laminated structure 层状结构
laminated(layered) clay 层状黏土
land caisson 陆上沉井
land drainage 地面排水
land subsidence 地面下沉，地面沉降
land upheaval 地面隆起
landfill 填土，填筑
landform 地貌，地形
landform element 地形单元

landform unit 地貌单元
landing runway 降落跑道
land-leveling 土地平整
landslide(slope failure) 滑坡
landslide correction 滑坡整治
landslide effects 滑坡影响
landslide mass 滑坡体
landslide prevention skirting 护脚
Laplacian operator 拉普拉斯算子
large bore 大直径钻孔
large deflection theory 大挠度理论
large region yield 大面积屈服
large scale field test 大规模野外试验，大型现场试验
large scale soil test 大型土工试验
large strain 大应变
large strain amplitude 大应变幅
large-diameter double tube 大直径双层取芯管
large-diameter drill rig 大直径钻机
large-diameter drilling machine 大孔径钻机
large-scale model 大比例模型
large-scale soil map 大比例尺土壤图
latent instability 潜在不稳定性
latent-water well 潜水井
lateral bearing capacity of piles 桩的横向承载力
lateral buckling 侧向压屈
lateral compression 侧压力，侧向挤压
lateral compression test 旁压试验
lateral confining pressure 侧限压力
lateral cutting 侧向开挖
lateral deformation 侧向变形，横向变形
lateral displacement 侧向位移
lateral drainage ability 侧向排水能力
lateral earth pressure 侧向土压力

lateral expansion 侧向膨胀
lateral force 横向力
lateral isotropy 横观各向同性体
lateral load 横向荷载
lateral load test 横向载荷试验
lateral loading test of pile 桩的横向载荷试验
lateral modulus of subgrade reaction 侧向地基反力系数
lateral movement stake 侧向移动标桩
lateral pile load test 桩侧向荷载试验
lateral pressure 侧压力
lateral pressure apparatus 侧压仪
lateral pressure coefficient 侧压力系数
lateral restraint 侧向约束
lateral restraint reinforcement 侧限加固
lateral rigidity 侧向刚度
lateral scouring 侧向冲刷
lateral shrinkage 侧向收缩
lateral slide 侧向滑动
lateral spread 侧向扩展
lateral squeezing-out of soft soil 软土的侧向挤出
lateral stability 侧向稳定性,横向稳定性
lateral strain 侧向应变
lateral strain indicator 侧向应变指示器
lateral stress 侧向应力
lateral support 横向支撑
lateral torsional buckling 侧扭屈曲
lateral underpinning 侧向托换
lateral yield 侧向屈服
laterite 红土
lateritic soil 砖红壤性土
laterization 红土化作用
lattice beam 格构梁
lattice charge 晶格电荷

lattice column　格构柱
lattice structure　晶格结构
lava　熔岩
lava ash　熔岩灰，火山灰
lava flow　熔岩流
lava tube　熔岩通道
law of conservation of energy　能量守恒定律
law of similarity　相似定律
lay-down thickness　松铺厚度
layer construction　分层铺筑
layer structure　层状结构
layered clay　层状黏土
layered media　层状介质
layered soil　成层土
layered strata　层状地层
layered system　层状体系
layerwise summation method　分层总和法
laying depth　埋置深度
layout of reinforcement　加固方案
leachates　滤出物
leaching　渗溶作用，淋溶
leaching effect　淋溶效应
leaching well　渗水井
lead　导架（打桩）
leader　落水管
leading pile　导桩
leading trench of diaphragm wall　连续墙导沟
leakage factor　渗漏系数
leakage loss　渗漏损失
leakage of mortar　漏浆
leakage of water　渗漏水
leakage path　渗径
leaking well　渗漏井
leakproof　防漏的

leaky foundation	漏水地基
lean clay	贫黏土，低塑性黏土
least-square method	最小二乘法
leck	硬黏土
Leda clay	来达黏土
length height ratio	长高比
length of drill rod	钻杆长度
length of restraint	嵌固长度
length to diameter ratio	长径比
lens	透镜体
levee	堤
levee revetment	堤面护坡
level of foundation	基础标高
level of safety	安全度
leveling	水准测量
levitating force	浮力
lift thickness	铺土厚度
light soils	轻质土
light sounding test blow count	轻便触探试验锤击数，N_{10}
lignite	褐煤
ligno-chrome gel	木铬胶
like-grained	单一粒径的，颗粒均匀的
lime clay	灰质黏土
lime column	石灰柱
lime compaction pile	石灰桩挤密
lime concretion	钙质结核
lime deep mixing method	石灰系深层搅拌法
lime earth concrete wall	三合土墙
lime earth rammed	灰土夯实
lime mortar	石灰砂浆
lime pile	石灰桩
lime pile method	石灰桩法
lime reactivity	石灰活化性

lime sink 落水洞
lime soil 石灰土
lime stabilization 石灰加固
lime stabilized soil 石灰稳定土
lime treated soil 灰土
lime-soil column 灰土井柱
lime-soil compaction pile 灰土挤密桩
lime-soil cushion 灰土垫层
lime-soil foundation 灰土基础
lime-soil pile 灰土桩
limestone 石灰岩
limestone cave 石灰岩溶洞
limit analysis 极限分析
limit design 极限设计
limit equilibrium method 极限平衡法
limit of error 误差限度
limit of liquidity 液限
limit of plasticity 塑限
limit of stability 稳定极限
limit of yielding 屈服极限
limit pressure 极限压力
limit state 极限状态
limit state design method 极限状态设计法
limit strength 极限强度
limit value 极限值
limiting equilibrium mechanics 极限平衡力学
limiting grain size 最大粒度
limiting load 极限荷载
limiting point instability 极值点失稳
limiting resistance 极限阻力
limit-load of instability 失稳极限荷载
limonite 褐铁矿
line load 线荷载
line of creep 蠕动线

line of saturation 浸润线，饱和线
line of sliding 滑动线
linear displacement 线位移
linear elastic analysis method 线弹性分析法
linear elastic model of soil 土的线弹性模型
linear expansion 线胀率
linear load 线荷载
linear shrinkage ratio 线缩率
linear strain 线应变
linear strain ratio 线应变率
linear variable differential transformer 线性可变差动变压器
linear viscoelastic model 线性黏弹性模型
linearity 线性
lined of excavation 开挖线
lined tunnel 衬砌隧洞
lining 衬砌
lining with column-typed sidewalls 柱式边墙衬砌
lining with continuous-arched sidewalls 连拱墙衬砌
linkage 键结，连接
linking beam of foundation 基础连系梁
liparite 流纹岩
liquefaction 液化
liquefaction defence measures 抗液化措施
liquefaction failure 液化破坏
liquefaction index 液化指数
liquefaction of sand 砂土液化
liquefaction potential 液化势
liquefaction strength 抗液化强度，液化强度
liquefaction zone 液化区
liquid limit 液限
liquid limit apparatus 液限仪

liquid limit device 液限仪
liquid limit test 液限试验
liquid phase 液相
liquid state 液态
liquidity index 液性指数
liquid-plastic limit combined device
　　　液塑限联合测定仪
lithification 岩化作用，石化作用
lithology 岩石学，岩性学
lithosol 石质土
lithosphere 岩石圈
littoral deposit 滨海沉积
littoral zone 滨海带，潮汐带
live load 活荷载
lixiviation 溶滤
load at first crack 开裂荷载
load at rupture 破坏荷载
load bearing capacity 承载力
load bearing layer 持力层
load bearing stratum 持力层
load cell 压力盒
load coefficient 荷载系数
load combination 荷载组合
load deflection curve 荷载挠度曲线
load distribution angle 基础刚性角
load effect combination 荷载效应组合
load factor 荷载系数
load factor method 荷载系数法
load gauge 测力计
load increment 荷载增量
load increment ratio 荷载增量比
load peak 荷载峰值
load ring 量力环
load stage 荷载阶段，加荷时期

load test on pile 桩的载荷试验
load transfer function method of pile
　　桩的荷载传递函数法
load transfer mechanism 荷载传递机理
load transfer method 荷载传递法
load-bearing capacity 承载能力
load-bearing test 承载试验，载荷试验
load-displacement curve 荷载-位移曲线
loaded area 承载面积
loaded filter 压载滤水体
loader 装载机
loading and unloading 加载与卸载
loading berm 反压护道，加压戗台
loading duration 荷载历时
loading frame 荷载架
loading function 荷载函数
loading history 荷载历史
loading in increment 分级加载
loading intensity 加载强度
loading method 加载方式
loading method of initial displacement
　　初位移加载法
loading method of initial velocity 初速度加载法
loading plate 承载板
loading plate test 平板载荷试验
loading platform 载荷台
loading sensing cell 荷载传感器
loading test 载荷试验
loading transfer 荷载传递
load-settlement curve 荷载-沉降曲线
loam 壤土，垆姆，亚黏土
local bearing area 局部承压面积
local bearing capacity 局部受压承载力
local buckling 局部屈曲

local erosion 局部冲刷
local flexure 局部弯曲
local hardening 局部硬化
local pneumatic process 局部气压法
local shear failure 局部剪切破坏
local side friction 局部侧摩阻
locking force 锁紧力
Lode's angle 罗德角
Lode's number 罗德数
Lode's parameter 罗德参数
loess 黄土
loess collapsibility test 黄土湿陷试验
loess collapsing 黄土湿陷
loess concretion 黄土结核，礓石
loess plateau 黄土高原
loessial soils 黄土类土
loess-like soil 黄土状土
log 钻孔柱状图，测井记录
log of bore hole 钻孔记录，钻孔柱状图
log scale 对数比例尺
logarithm of time fitting method 时间对数拟合法
logarithmic decrement 对数减量
logarithmic spiral 对数螺线
loglog plot 双对数坐标图
London clay 伦敦黏土
long dolly 送桩
long hole blasting 深孔爆破
long hole drilling 深孔钻进
long hole grouting 深孔灌浆
long run test 长期试验
long term loading test 长期荷载试验
long term stability of slope 边坡长期稳定
long-hole method 深孔爆破法
longitudinal bending factor 纵向弯曲系数

longitudinal buckling 纵向弯曲
longitudinal crack 纵向裂缝
longitudinal dike 顺坝
longitudinal displacement 纵向位移
longitudinal permeability 纵向透水性
longitudinal profile 纵剖面
longitudinal rigidity 纵向刚度
longitudinal wave 纵波，P波
long-term behavior 长期性状
long-term creep 长期蠕变
long-term modulus 长期模量
long-term stability 长期稳定性
long-time loading 长期荷载
long-time strength of rock 岩石长期强度
loose alluvium 松散冲积层
loose blasting 松动爆破
loose coefficient of soil 土壤松散系数
loose foundation 松散地基
loose pressure of surrounding rock 围岩松动压力
loose rock mass 松散岩体
loose sand 松砂
loose soil 松散土
loosely bound water 弱结合水
looseness of soil 土壤松散度
loosening pressure 松动压力
loosening zone of surrounding rock 围岩松动区
Los Angeles abrasion test 洛杉矶磨损试验
loss due to anchorage take-up 锚固损失
loss of creeping 徐变损失
loss of head 水头损失
loss of pressure 压力损失
loss of prestress 预应力损失
loss of soil 水土流失
loss of weight 重量损失，失重

Love wave 乐甫波，Q 波
low pile cap 低桩承台
low strength cement 低强度水泥
lower bound theorem 下限定理
lower plastic limit 塑限下限
lowest erosion line 最低冲刷线
Loyang(Luoyang)spoon 洛阳铲
Lugeon test 吕荣试验
Lugeon unit 吕荣单位
lump soil 块土
lumped mass 集总质量
lumped parameter method 集总参数法
lunar soil 月球土壤
lunar soil mechanics 月球土力学
lysimeter 测渗计，溶度计

M

m coefficient method m 法（横向受荷桩分析）
macadam 碎石，碎石路
macadam foundation 碎石基础
macadam pavement 碎石路面
machine foundation 机器基础
machine rock drill 岩石钻机
macrocrack 宽裂缝
macrofabric 粗组构，宏观组构
macropore 大孔隙
macropores coefficient 大孔隙系数
macroporous soil 大孔性土，黄土
macroporous structure 大孔结构
macroscopic granular 粗颗粒的
macroseism 强震
macrostructure 宏观结构

macrovoid ratio 大孔隙比
made ground 填方,填土
magma 岩浆
magmatic rock 岩浆岩
magnesian limestone 镁质石灰岩
magnesium octahedron 镁氧八面体
magnetic probe extensometer 磁性沉降计
magnetic prospecting 磁性勘探
magnetite 磁铁矿
magnetoelectric pile driver 电磁打桩机
magnetoelectric pile hammer 电磁桩锤
magnetometer 磁力仪
magnification factor 放大系数
magnitude 震级
magnitude chart 震级图
magnitude determination 震级测定
magnitude distribution 震级分布
magnitude of load 荷载大小
main arch ring 主拱圈
maintained load test 维持荷载法
major defect 主要缺陷
major principal plane 大主平面
major principal strain 大主应变
major principal stress 大主应力
Malan loess 马兰黄土
Mandel-Cryer effect 曼代尔-克雷尔效应
mandrel 芯棒(桩工)
manhole 检查井
man-made fiber 人造纤维
man-made seismic hazards 人为震害
man-made slope 人工边坡
manometer 压力计
manostat 稳压器
mantle of soil 土的表层,表皮土

mantle rock 风化层，土被，表皮土
manual boring 人力钻探
manual sampling 人工取样
manually excavated cast-in-place pile 挖孔灌注桩
map of burial depth of bedrock 基岩埋深图
marble 大理石
Marchetti's flat dylatometer 马氏扁式松胀仪
marginal value 临界值
marine clay 海相黏土
marine deposit 海相沉积
marine geotechnology 海洋岩土工程学
marine rock 海成岩石
marine sampling 海洋取样
marine soil 海积土
marine soil mechanics 海洋土力学
marl 泥灰岩
marly clay 泥灰质黏土
marsh 沼泽
marsh soil 沼泽土
marshland 沼泽地
masonry 砌体工程，砖石工程，砌体，砌筑，砖石建筑
masonry retaining wall 砌体挡墙
mass density 质量密度
mass force 体积力
mass movement 整体移动
mass properties 整体特性
mass ratio 质量比
massive and rigid structure 大体积刚性结构
massive foundation 大块式基础
massive rock 块状岩
mass-spring-dashpot system 质量-弹簧-阻尼器体系
mat foundation 筏形基础

material damping 材料阻尼
material non-linearity 材料非线性
material properties 材料特性
material retained 筛余
mathematical model 数学模型
matric potential 广义毛管势
matrix 母岩，矩阵
matrix suction 基质吸力
mattress 垫层
maximum aggregate size 集料最大粒径
maximum consolidation pressure 最大固结压力
maximum depth of frozen ground 最大冻土深度
maximum depth of plastic zone 塑性区最大深度
maximum displacement 最大位移
maximum drilling depth 最大成孔深度
maximum dry density 最大干密度
maximum dry unit weight 最大干重度
maximum extracting force on pile 最大拔桩力
maximum molecular water content 最大分子吸水量
maximum particle size 最大粒径
maximum past pressure 最大前期压力
maximum permissible load 最大容许荷载
maximum pressure on pile 最大压桩力
maximum principal stress 最大主应力
maximum saturation 最大饱和度
maximum shear theory 最大剪应力理论
maximum void ratio 最大孔隙比
Maxwell model 麦克斯韦尔模型
mean diameter 平均粒径
mean effective stress 平均有效应力
mean principal stress 平均主应力
mean sea level 平均海平面
mean specific gravity 平均比重

measured geological map 测定地质图
measuring device 量测仪器
measuring error 量测误差
measuring glass 玻璃量杯
measuring range 量程
mechanical analysis 粒径分析
mechanical analysis curve 粒径分析曲线
mechanical composition 颗粒级配
mechanical impedance 机械阻抗
mechanical properties 力学性质
mechanical similarity criterion 力学相似性
mechanical stabilization 改善级配加固法
mechanical weathering 机械风化
mechanics of granular media 散体力学
mechanics of jointed rock 节理岩石力学
mechanics of mixture 混合体力学
mechanism 机理
mechanism for soil liquefaction 土液化机理
mechanism of rock mass deformation 岩体变形机制
median(medium)diameter 中值粒径
medical lock 医疗匣(沉箱)
medium consistency 中等稠度
medium grained soil 中粒土
medium sand 中砂
Medvedev scale 麦德维捷夫地震烈度表
megascopic method 肉眼认别法
megavarve 粗纹泥
mellow soil 松软土
membrane 橡皮膜(三轴试验)
membrane compliance 橡皮膜顺变性
membrane correction 橡皮膜校正
membrane method 薄膜法
membrane penetration effect 橡皮膜嵌入效应

membrane potential 膜电位
membrane type fabric support 膜式织物支撑
membrane type pressure gauges 薄膜式压力盒
Menard pressure meter 梅纳旁压仪
meniscus 弯液面
Mercalli scale 麦卡利地震烈度表
mercury gauge 汞测压计
mesh 筛孔，网
mesh analysis 筛分析
mesh reinforcement 钢筋网
Mesozoic era 中生代
metal sheet pile 钢板桩
metallic bond 金属键
metamorphic petrology 变质岩石学
metamorphic rock 变质岩
metastable soils 亚稳土
metastable structure 亚稳结构
meteoric water 雨水
meteorological water 大气水
method for settlement calculation 沉降计算法
method of drained sinking 排水下沉法
method of floating open caisson 浮运沉井法
method of forced vibration 强迫振动法
method of grading mud-making clay
   黏土分级评价法
method of sampling 取样法
method of slice 条分法
method of slurry direct circulation 泥浆正循环法
method of slurry reverse circulation 泥浆反循
                环法
method of superposition 叠加法
method of undrained sinking 不排水下沉法
Meyerhof's formula 梅耶霍夫承载力公式
mica 云母

mica-basalt 云母玄武岩
micaceous shale 云母页岩
mica-marble 云母大理石
mica-schist 云母片岩
micellae 胶粒
micro texture 微观组织
micro-aggregate 微团粒
microbiological treatment 微生物处理
micro-crack 微裂纹
microfabric 微组构
microfissure 微裂隙
microfossil 微化石
micrometer 测微计
micro-pile 小型桩
microseism 微震
microstructure 微观结构
micro-texture 微观组织
microwave moisture apparatus 微波含水量仪
middle water 层间水
mil 密尔(单位)
mild clay 软黏土
Mindlin's solution 明德林解答
mine refuse 矿渣
mine tailings dam 尾矿坝
mineral phase 矿物相
mineral skeleton 矿物骨架
mineralized degree of groundwater 地下水矿化度
mineralized water 矿化水
mineralogical analysis 矿物分析
mineralogical composition 矿物成分
mineralogy 矿物学
mini pile 微型桩
minimum embedded depth of foundation
　　　基础最小埋深

minimum penetration　最小贯入度
minimum pile spacing　桩的最小间距
minimum safety stability　最小安全稳定性
minimum void ratio　最小孔隙比
mining subsidence　采空塌陷
minor geologic details　次要地质细节
minor principal plane　小主平面
minor principal strain　小主应变
minor principal stress　小主应力
mire　淤泥
miscellaneous fill　杂填土
Mises yield condition　米赛斯屈服条件
mixed sedimentary rock　混合沉积岩
Mixed-in-place　现场拌合
mixed-in-place pile　拌合桩，就地搅拌桩
mobile ground　流沙层
mode of deposition　沉积条件
mode of failure　破坏模式
mode of vibration　振型
model law　模型律
model of idealized elastic-plastic deformation
　　　理想弹塑性变形模型
model of idealized rigid-plastic deformation
　　　理想刚塑性变形模型
model of mechanics　力学模型
model scale　模型比例尺
model test　模型试验
modeling gradation　模拟级配
moderate sea deposit　半深海沉积
moderately soluble salt test　中溶盐试验
modes of shear failure for subsoil
　　　地基剪切破坏模式
modified Griffith's criterion
　　　修正的格里菲斯准则

modified Mercalli scale 修正麦卡利地震烈度表
modular ratio effect 模量比效应
modulus of compressibility(compression) 压缩模量
modulus of deformation 变形模量
modulus of dilatation 膨胀系数，膨胀模量
modulus of elasticity 弹性模量，杨氏模量
modulus of elasticity in shear 剪切弹性模量
modulus of elongation 拉伸模量
modulus of linear deformation 线性变形模量
modulus of plasticity 塑性模量
modulus of pressure meter 旁压仪模量
modulus of resilience 回弹模量
modulus of rigidity 刚性模量，抗剪模量
modulus of section 截面模量
modulus of shear deformation 剪切变形模量
modulus of shear resilience 剪切回弹模量
modulus of subgrade reaction 地基反力系数
modulus ratio 模量比
Mohr scale of hardness 摩尔硬度计
Mohr strength envelope 摩尔强度包线
Mohr's envelope 摩尔包线，强度包线
Mohr-Coulomb criterion 摩尔-库仑准则
Mohr-Coulomb law 摩尔-库仑定律
Mohr-Coulomb soil model 摩尔-库仑模型
Mohr-Coulomb theory 摩尔-库仑理论
Mohr's circle 摩尔圆
moist rodding 湿捣法
moist tamping 湿击法
moisture 湿度
moisture barrier 防潮层
moisture can 含水量盒，土盒
moisture content 含水量

moisture content of natural 天然含水量
moisture equivalent 含水当量
moisture expansion 湿膨胀
moisture index 水分指数，土壤含水量指数，含水指数
moisture migration 水分迁移
moisture probe 同位素含水量探测仪
moisture retention 保水性
moisture room 保湿室
moisture-density curve 击实曲线
moisture-density nuclear gauge 含水量-密度核子测定仪
moisture-density test 击实试验
mole drain 地下排水沟
mole drainage 地下排水工程
molecular attraction 分子引力
molecular bond 分子键
molecular cohesion 分子黏聚力
molecular force 分子力
mole-plough 开沟犁
mollisol 松软土
moment of inertia 惯性矩
moment of inertia of cross-section 截面惯性矩
monitor 检测器，喷水枪
monitoring 监测
monitoring of landslide 滑坡监测
monitoring of pore-water pressure 孔隙水压力监测
monitoring of settlement and deformation 沉降变形监测
monitoring of surrounding rock deformation of tunnel 洞室围岩变形监测
monkey 打桩锤
monolithic retaining walls 整体式挡土墙

monotube pile 单管桩
monovalent ion 单价离子
montmorillonite 蒙脱石
moor soil 泥炭土,沼泽土
moor peat 沼煤
moorstone 花岗石
morainal deposit 冰碛沉积
moraine 冰碛
morass 沼泽
Morgenstern method of slope stability 摩根斯坦法(边坡稳定分析)
mortar 砂浆
mortar bound arch 浆砌拱圈
mortar grouting method 压浆法
mother rock 母岩
mottled sandstone 杂色砂岩
mountain peat 山地泥炭
muck 淤泥,腐殖土,废土石
muck foundation 淤泥地基
mucky soil 淤泥质土
mud 泥,泥浆
mud analysis log 泥浆分析记录
mud and rock flow 泥石流
mud avalanche 泥石流
mud balance 泥浆天平
mud cake 泥皮
mud circulation 泥浆循环
mud disposal 废泥浆
mud ditch 泥浆槽
mud formula 泥浆配方
mud grouting 灌泥浆
mud injection 泥浆灌注
mud jack 压浆泵
mud jack method 压浆法

mud jacking 压泥浆
mud layer 泥浆层
mud materials 泥浆材料
mud pump 泥浆泵，吸泥泵
mud pumping 翻浆冒泥
mud spouts 喷水冒砂
mud trap 泥浆池
mud treatment 泥浆处理
mud treatment plant 泥浆处理装置
mud wave 泥滑
mudding 翻浆
mud-setting pit 泥浆沉淀池
mudstone 泥岩
muffler piling 消声打桩法
mulching soil 覆盖土
multi-anchorage system 多层锚定系统
multi-anchored wall 梯式加筋锚定墙
multiaxial tensile test 多轴拉伸试验
multi-defective pile 多缺陷桩
multi-dimensional consolidation 多维固结
multi-directional shear 多向剪切
multilayer semi-infinite solid 多层半无限体
multi-membrane 多层薄膜
multiphase 多相的
multiple injection recorder 多孔灌浆记录仪
multiple line curtain 多排帷幕
multiple piezometer 多头测压管
multiple pile 群桩
multiple point extensometer 多点应变计
multiple reflection 多次反射
multiple stress concentration 多重应力集中
multiple use 综合利用
multi-purpose triaxial test apparatus
    多能三轴仪

multi-reinforcement 多层加强
multi-section pile 多节桩
multi-stage drainage 多级排水
multi-stage triaxial compression test
　　　　多级三轴压缩试验
multi-stage well point system 多级井点系统
multi-well open caisson 多排孔沉井
municipal landfill 城市填土
municipal waste 城市废料
muscovite 白云母
muscovite-granite 白云母花岗石
myckle 软黏土
mylonite 糜棱岩

N

narrow gradation 均匀级配
native soil 天然土
natural angle of repose 天然休止角
natural arch 天然拱
natural base(foundation, ground)
　　　　天然地基，底土
natural boundary conditions 自然边界条件
natural building materials 天然建筑材料
natural conservation 自然(资源)保护
natural consistency test 天然稠度试验
natural drainage 天然排水
natural drying 自然干燥
natural environment 自然环境
natural erosion 自然侵蚀
natural freezing(frozen)method 天然冻结法
natural frequency 固有频率
natural frequency of foundation 基础自振频率

natural impregnation 自然浸渍
natural load-transmitting arch 天然卸载拱
natural moisture content 天然含水量
natural period 固有周期
natural sand 天然砂
natural sedimentation 自然沉淀
natural slope 天然坡
natural slope angle of soil 土壤自然坡度角
natural soft clay 天然软黏土
natural soil 天然土
natural soil drainage 土壤自然排水
natural strata 天然地层
natural vibration 固有振动
natural void ratio 天然孔隙比
natural water content 天然含水量
navvy 挖掘机
near-mesh grain 分界粒度颗粒
nearshore 近岸，近滨
nearshore structure 近海结构物
neat cement grout 净水泥浆
necking 颈缩
necking phenomena 颈缩现象
needle hole test 针孔试验
needle penetrometer 针式贯入仪
needle-punched geotextile 针刺土工织物
needle-punched process 针刺法
needle-shaped structure 针状构造
negative friction pile 负摩擦桩
negative mantle friction 负摩擦力(桩工)
negative pore water pressure
　　　　负孔隙(水)压力
negative skin friction of pile 桩的负摩阻力
neighbouring pile 邻桩
Neogene period 晚第三纪

neo-loess 新黄土
Neozoic era 新生代
net contact(foundation)pressure 基底附加压力
net loading intensity 净荷载强度
neutral point of pile 桩的中性点
neutral pressure 中性压力
neutral soil 中性土壤
neutral stress 中性应力
neutron logging 中子测井
New Austrian Tunnelling Method(NATM) 新奥法
new stone age 新石器时代
Newmark's(influence)chart 纽马克感应图
Newton model 牛顿模型
Newton's theory of collision 牛顿碰撞理论
Newtonian fluid 牛顿流体
Newtonian liquid 牛顿液体
nitosol 强风化黏磐土
no conservative force 非保守力
nodal displacement 节点位移
nodal force 节点力
no-load saturation curve 无载饱和曲线
non cohesive soil 无黏性土
non plastic 非塑性的
non plastic soil 无塑性土
non-artesian ground water 非自流地下水
non-artesian well 非自流井
non-associative flow rule 非关联流动法则
nonaxial load 偏心荷载
non-circular analysis 非圆弧分析法
non-clay mineral 非黏土矿物
non-cohesive 非黏性的
non-cohesive soil 松散岩土,非粘结性土
non-collapsing soil 非湿陷性土

non-confined compression test 无侧限压缩试验
noncore drilling 无岩芯钻进，不取芯钻进
noncrystalline clay mineral 非晶质黏土矿物
non-destructive detection 无损探伤
non-destructive techniques 无损探测技术
non-destructive test 非破损试验
non-destructive testing(NDT) 非破坏性测试，无损探伤试验
non-displacement pile 钻孔灌注桩，钻孔桩
nongraded mix 非级配混合料
nonhomogeneity 不均匀性
nonhomogeneous soil 非均质土
nonlinear analysis 非线性分析
nonlinear boundary condition 非线性边界条件
nonlinear constraints 非线性约束
nonlinear correlation 非线性相关
nonlinear elastic behavior 非线性弹性性状
nonlinear elastic model 非线性弹性模型
nonlinear elasticity theory 非线性弹性理论
nonlinear finite element analysis 非线性有限元分析
nonlinear large deflection theory of stability 非线性大挠度稳定理论
nonlinear material problem 材料非线性问题
nonlinear model 非线性模型
nonlinear theory 非线性理论
nonlinear viscoelastic model 非线性黏弹性模型
nonlinearity 非线性
non-particulate grout 非颗粒性灌浆
non-pressure flow 无压流
non-scouring velocity 止冲流速
non-seismic region 非地震区
non-slaking clay 非湿化性黏土

non-stationary flow 非稳定流
non-tectonic fault 非构造断层
non-tectonic ground crack 非构造地裂缝
non-tectonic joint 非构造节理
nonuniform flow 非均匀流
nonuniform grading 不均匀级配
nonuniform settlement 不均匀沉降
nonwoven geotextile 无纺土工织物
non-yielding retaining wall 刚性挡土墙
normal compressive stress 法向压应力
normal consolidation line 正常固结线
normal distribution 正态分布
normal force 法向力
normal pressure 法向压力
normal stiffness 法向刚度
normal strain 法向应变
normal stress 法向应力
normal-circulation rotary drilling 正循环旋转钻进
normality condition 正交条件
normality law 正态分布律
normalization 归一化，标准化，规格化
normally consolidated(consolidation) soil 正常固结土
North Dakota cone test 北达科他州圆锥试验
Norwegian clay 挪威黏土
nose pollution 噪声污染
nuclear densometer 核子密度仪
nuclear soil moisture meter 核子土壤湿度计
nuclear waste disposal 核废料处理
null indicator 零位指示器
number of blows 锤击数
number of piles 桩数（群桩）
number of roller passes 碾压遍数

numerical analysis 数值分析
numerical geomechanics 数值岩土力学
numerical integration 数值积分
nylon 尼龙

O

oblique joint 斜节理
obliquity 倾斜度
observation borehole 观测井
observation of ground stress 地应力观测
observation of surrounding rock deformation
　　　围岩变形观测
occurrence 产状
occurrence condition of groundwater
　　　地下水产出条件
ocean engineering 海洋工程学
ocean current 洋流
ocean soil 海洋土
ocean structure 海洋结构物
oceanic deposit 海洋沉积
octahedral layer 八面体层
octahedral normal stress 八面体正应力
octahedral shearing stress 八面体剪应力
oedometer 固结仪
oedometer test 固结试验
oedometric modulus 侧限压缩模量
oedotriaxial test 固结仪三轴仪联合试验法
off-angle drilling 钻斜孔法
offset of foundation 基础襟边
offset tangent modulus 补偿切线模量
offshore boring 海底钻探
offshore deposit 近海沉积

offshore piles 近海桩
offshore platform 近海台地，海上平台
offshore soil mechanics 近海土力学
offshore structure 近海结构物
offshore terrace 近海阶地
offshore well drilling platform 近海钻井平台
oil soil 古土壤
old loess 老黄土
one dimensional consolidation 一维固结
one-shot method 单液法（化学灌浆）
one-side water pressure 单侧水压力
one-way drainage 单面排水
on-job laboratory 工地实验室
onshore piles 岸上桩
on-site identification of soil and rock
　　　土石场现场鉴别法
on-site investigation 现场调查
on-site work 现场工作
ooze 海底淤泥
oozing 渗漏
open area ratio 开孔面积比
open caisson 沉井，开口沉箱
open caisson foundation 沉井基础
open caisson method 沉井法
open caisson sinking by suction dredge
　　　吸泥下沉沉井
open caisson wall 沉井壁
open caisson with a tailored cutting edge
　　　高低刃脚沉井
open caisson with cross walls 格墙沉井
open caisson with two shells of wire-mesh cement
　　　双壁钢丝网水泥沉井
open cut method 明挖法
open cut foundation 明挖基础

open cut tunnel 明挖隧道，明洞
open drain 明沟
open excavation 明挖
open fissure 开口裂隙
open grain structure 大孔结构
open sand 松砂
open sheeting 间隔挡板
open tube diesel hammer 开管式柴油锤
open water 地表水
open work gravel 架空砾石层
open-end pipe pile 开口管桩
open-end steel pile 开口钢桩
open-ended grouting 敞口灌浆
open-hole drilling 无套管钻探
openlagging 间隔挡土板
operating rules 操作规程
optimal gradation 最佳级配
optimal limit design 极限优化设计
optimal rolling depth 最佳压实深度
optimization for design 优化设计
optimization model 优化模型
optimum approximation 最佳逼近
optimum density 最佳密度
optimum depth 最佳覆盖厚度
optimum fine aggregate percentage
　　　最佳含砂率
optimum moisture(water)content 最优含水量
optimum reinforcement layer number
　　　最优加筋层数
order of magnitude 数量级
ordinary deposited soil 一般堆积土
ordinary masonry 毛石圬工
ordinary water level 常水位
Ordovician period 奥陶纪

organic clay 有机黏土
organic compound 有机化合物
organic content 有机质含量
organic environment 有机环境
organic ion 有机离子
organic matter 有机质
organic soil 有机质土
organic sphere 生物界
organogenous sediments 有机沉淀物
orientation 定向，定位
origin cohesion 初始黏聚力
original ground 原地面
original joint 原生节理
original mineral 原生矿物
original pressure 原始地层压力
original riverbed 原河床
original rock 原岩，基岩
orterde 硬土层
orthoclase 正长石
orthogonal joint 横节理
orthometamorphite 正变质岩
orthotropic 正交各向异性的
orthotropic elasticity 正交各向异性弹性体
orthotropic material 正交各向异性材料
orthotropic particle 正交各向异性颗粒
orthotropy 正交各向异性
ortstein 褐铁矿，灰质壳
oscillating screen 振动筛
oscillation pickup 拾振器
oscillograph 示波器
oscillometer 示波仪
osmose movement 渗透运动
osmosis 渗透，渗析
osmotic phenomenon 渗透现象

osmotic potential 渗透势
osmotic pressure 渗透压力
osmotic pressure method 渗透压法
osmotic suction 渗透吸力
osmotic water 渗透水
Ottawa sand 渥太华砂
out and cover method 明挖法
outcrop 露头
outcrop of water 水流出逸口
outlet velocity 出口流速
outside clearance ratio 外间隙比(取样器)
outwash 冰水沉积，冲刷
outwash plain 冰川沉积平原
oven-dried soil 烘干土
over coarse-grained soil 巨粒土
over consolidation ratio(OCR) 超固结比
over load test 超载试验
over stretching 超张拉
overall buckling 整体屈曲
overall dimension 总尺寸
overall flexure 整体弯曲
overall inclination 整体倾斜
overall site layout 施工现场总平面图
overall stability 整体稳定
overall stress 总应力
overall unstability 整体失稳
overbreak 超挖
overburden grouting 覆盖层灌浆
overburden layer 覆盖层
overburden pressure 覆盖压力
overcharge 超载
overcompacted clay 超压实黏土
overcompaction 过度压实
overconsolidated clay 超固结黏土

overconsolidated soil 超固结土
overconsolidation 超固结
overconsolidation soil 超固结土
overcoring 套钻
overdamping 超阻尼
overdesign 保守设计
overdriven pile 超打桩
overdriving 超打(打桩)
overexcavate 超挖
overfill 超填
overhanging bank 陡岸，悬岸
overlapping pile 叠置桩
overlaying rock 上覆岩石
overload 超载
overload capacity 超载能力
overload characteristics 超载特性
overlying bed 覆盖层，上覆层
overlying crust 硬壳
overlying strata 上覆地层
overpumping 过量抽水
over-rolling 过度碾压
oversaturation 过饱和
oversize particle 超径颗粒
oversized materials 超径材料
overstrain 超限应变
overstress 超限应力
overstress load 超应力荷载
overstressed rock mass 超应力岩体
overturning force 倾覆力
overturning moment 倾覆力矩
overturning stability 倾覆稳定性
overweight 超重
oxbow lake 牛轭湖
oxidation 氧化

oxide 氧化物

P

P-Δ effect   P-Δ 效应
packed(fabri-enclosed)drain   袋装砂井
packer   灌浆孔塞，夯具，捣实器
packer grouting   栓塞灌浆（法），分段灌浆
packer test   分段栓塞水压试验
packing   填料，衬料，衬垫（打桩）
pad foundation   垫式基础，独立基础
padding   填料，填塞，大石块，衬垫，衬底织物，背衬
paint base   油漆打底，底漆
paint filler   油漆底层
Palaeozoic era   古生代
paleo clay   老黏土
paleo-deposited soil   老堆积土
Paleogene period   古近纪
pan formation   硬土层
pan soil   硬土
paper clay   薄层黏土
paper drain   纸板排水
paraffin   石蜡
paraffin wax   粗石蜡
parallel grading method   平行级配法
parallel section method   平行断面法
parallelogram of forces   力的平行四边形
parameter   参数
parameter curve   参数曲线
parameter detection   参数检测
parameter equation   参数方程
parameter estimation   参数估计

parameter identification 参数辨识
parameter inversion 参数反演
parameter of drilling and blasting 钻爆参数
parameter of saturation 饱和参数
parameter optimization 参数优化
parapet wall 防浪墙，女儿墙
parent rock 母岩
partial(safety)factor for action 作用分项系数
partial desiccation 半干燥，半脱水
partial erosion 局部冲刷
partial excavation 局部开挖
partial factor 分项系数
partial factor for load 荷载分项系数
partial factor for resistance 抗力分项系数
partial factor method 分项系数法
partial penetrating well 不完全渗水井
partial safety factor 部分安全系数(用于极限状态设计)
partial saturation 不完全饱和，非饱和，部分饱和
partial submergence 部分浸水
partially drained loading 部分排水加载
partially penetrated well 非完整井
partially saturated soil 部分饱和土
partially weathered zone 部分风化带
particle association 颗粒组合
particle cross-linking 颗粒交联
particle density 颗粒密度
particle diameter 颗粒直径
particle distribution 颗粒分布
particle dynamics 质点动力学
particle grading curve 颗粒级配曲线
particle shape 颗粒形状
particle size 颗粒大小

particle size accumulation curve 粒度累积曲线
particle size analysis 颗粒分析试验，粒径分析
particle size classification 粒度分级
particle size distribution 粒径分布
particle size factor 粒度系数
particle strength 颗粒强度
particle-size-distribution of soil 土的颗粒级配
particulate grouting 微粒灌浆
particulate mechanics 粒体力学
passive earth pressure 被动土压力
passive failure 被动破坏
passive isolation 消极隔振
passive pile 被动桩
passive Rankine pressure 被动朗肯土压力
passive Rankine zone 朗肯被动区
passive state of plastic equilibrium
　　　被动塑性平衡状态
path of loading 加载路径
path of percolation 渗透路径
path of seismic waves 地震波传播路径
pavement 路面
pavement pumping 路面抽水作用
pea gravel 小砾石，豆石
peak pore pressure ratio 峰值孔压比
peak shear strength 峰值抗剪强度
peak strain 峰值应变
peak strength 峰值强度
peak value 峰值
peat 泥炭
peat bog 泥炭沼泽
peaty soil 泥炭质土
pebble 卵石，鹅卵石
pedalfer 淋余土

pedestal 柱脚
pedestal pile 扩底桩
pedogenesis 成土作用
pedology 土壤学
pelagic deposit 深海沉积
pelite 泥质岩
pelitic structure 泥岩状结构
pelleter drier 颗粒干燥器
pellicular water 弱结合水,薄膜水
pendulum hammer(pile-driver) 摆式打桩机
penetrability 穿透性,渗透性
penetrating cone 触探锥,贯入锥
penetration characterisitics 渗透性能
penetration distance 渗透距离
penetration needle 贯入针
penetration of pile 沉桩
penetration of pile with hammer blow 锤击沉桩
penetration of slag 侵蚀
penetration of water 渗水
penetration per blow 每击贯入度(打桩)
penetration rate 贯入速率,穿透力
penetration resistance 贯入阻力
penetration resistant 抗渗性,抗渗的
penetration sounding 触探
penetration technique 贯入技术
penetration test 贯入试验,触探试验
penetrometer 稠度计,贯入度仪
penn stone 平扁石
perambulation 查勘,踏勘
percent compaction 压实系数
percent consolidation 固结度
percent fines 细粒含量百分率
percent of sample recovery 土样采取率
percent open area 孔眼面积百分数

percent saturation 饱和度
percentage air voids 含气率
percentage bulking 湿胀率
percentage by volume 体积百分比
percentage by weight 重量百分比
percentage of accumulated sieve residues 累计筛余百分数
percentage of bed-expansion 膨胀率
percentage of consolidation 固结百分率
percentage of core recovery 岩芯采取率
percentage of particle breakage 颗粒破碎率
percentage of sand 砂率
percentage of saturated water content 饱和含水率
percentage of void 空隙率
perched aquifer 上层滞水含水层
perched water 上层滞水
percolation 渗漏
percolation apparatus 渗透仪
percolation coefficient 渗透系数
percolation pit 渗坑
percolation rate 渗透速率
percolation test 渗滤试验,渗透试验
percussion and grabbling method 冲抓法
percussion borehole 冲击钻孔
percussion boring 冲击钻探
percussion drill 冲击钻机
percussion drilling 冲击钻进
percussion hand boring 人力冲击钻探
percussion table 振动台
perennial frozen soil 多年冻土
perfect elastic material 理想弹性材料
perfect plastic material 理想塑性材料
perfect plasticity 理想塑性
perforated casing 多孔套管

perforated drain pipe 多孔排水管
perforated stone 透水石
perimeter grouting 周边灌浆
period of vibration 振动周期
periodic load 周期荷载
periodical vibration 周期振动
periphery beam 圈梁
perlite 珍珠岩
permafrost 永冻土
permafrost aggradation 永冻土增厚
permafrost degradation 永冻层减薄
permafrost table 永冻土面
permanent bed 稳定河床
permanent deformation 永久变形，残余变形
permanent expansion 残余膨胀
permanent lining 永久支护
permanent load 永久荷载
permanent strengthening 永久性加固
permeability 渗透性
permeability coefficient 渗透系数
permeability meter 透气性测定仪
permeability of rock 岩石透水性
permeability of soil 土的渗透性
permeability reducing admixture 抗渗剂
permeability test 渗透试验
permeability to water 透水性
permeable blanket 透水铺盖
permeable grouting 渗入性灌浆
permeable joint 透水缝，渗透缝
permeable layer 透水层
permeable rock 透水岩层
permeable zone 渗透区
permeameter 渗透仪
permeation grouting 渗透灌浆

Permian period 二迭纪
permissible deflection 允许挠度
permissible load 容许荷载
permissible settlement 容许沉降量
permissible soil pressure 容许土压力
permissible stress 容许应力
permissible stress design method 容许应力设计法
permissible stress method 容许应力法
perpendicularity 垂直度
perpetually frozen soil 永久冻土
persorption 吸混作用(多孔性吸附)
perspex 有机玻璃
pervious bed 透水层
pervious blanket 透水垫层
pervious foundation 透水地基
pervious materials 透水材料
pervious zone 渗透区,透水带
perviousness 渗透性
petrification 石化作用
petrographic classification 岩石分类
petrographic composition 岩石组成,岩石成分
petrology 岩石学
petrotectnonics 岩石构造学
pH value pH值
phase angle 相角
phase difference 相位差
phase-plane 相平面
photographic inclinometer 摄影测斜仪
phreatic decline 地下水位下降
phreatic discharge 地下水流量,潜水涌出量
phreatic high 地下水上部含水层
phreatic line 浸润线,地下水位线
phreatic low 低地下水位,低潜水位
phreatic rise 地下水上升

phreatic surface 潜水面，地下水面
phreatic water 潜水，地下水
phreatic water level(surface) 潜水水位
phyllite 千枚岩
physical interaction 物理相互作用
physical landform 地形
physical model 物理模型
physical properties of rock 岩石的物理性质
physical properties of rock under water
　　　岩石的水理性质
physical similitude 物理相似
physical weathering 物理风化
physicomechnanical property of rock
　　　岩石的物理力学性质
physiography 地文学，地相学
picket 木桩，支柱
picnometer 比重瓶
piedmont 山麓
pier 桥墩
pier foundation 墩式基础
pierhead 防波堤堤头部
pierre perdue 水冲抛石
piezocone 可测孔隙水压力的触探头
piezocone test(CPTU) 孔压静力触探试验
piezometer 压力计，水压计
piezometer opening 测压孔
piezometer tip 水压测头
piezometer tube 测压管
piezometric head 测压管水头
piezometric level 测压管水位
piezometric regime 测压状态
piezometric surface 测压管水面
pile 桩
pile action 桩承作用

pile anchor 桩锚
pile band 桩箍
pile bearing 桩支承
pile bent 桩排架，桩桥台
pile bent pier 桩式桥台
pile body(shaft) 桩身，桩体
pile cap 桩帽，桩承台
pile cap as beams 梁式承台
pile cap as plate 板式承台
pile cap beam 桩顶连系梁
pile cap method 桩帽加固法
pile capacity 桩的承载力
pile capping beam 桩承台梁
pile casing 护筒
pile cluster 桩群
pile cofferdam 桩式围堰
pile core 桩芯
pile corrosion 桩的腐蚀
pile cross section 桩的截面
pile cushion 桩垫
pile cut-off 桩截水墙
pile drawer 拔桩机
pile driven formula 打桩公式
pile driver 打桩机
pile driver barge(boat) 打桩船
pile driver crane 打桩起重机
pile driver lead 打桩机导向柱
pile driving 打桩
pile driving analyser 打桩分析器
pile driving by vibration 振动打桩
pile driving efficiency 打桩效率
pile driving equipment 打桩设备
pile driving formula 打桩公式
pile driving frame 打桩架

pile driving machinery 桩工机械
pile driving platform 打桩平台
pile driving record 打桩记录
pile driving resistance 打桩阻力
pile driving stresses 打桩应力
pile driving test 打桩试验
pile end bearing 桩端承载力
pile extension 接桩
pile extraction 拔桩
pile extraction resistance 拔桩阻力
pile extractor 拔桩机
pile fabricating yard 制桩场
pile fender 护桩,防冲桩
pile flutter 桩的颤动
pile follower 送桩
pile footing 桩承基脚
pile force gauge 桩力计
pile foundation 桩基础
pile foundation with high cap 高承台基桩
pile foundation with low cap 低承台基桩
pile grid 桩承台
pile group 群桩
pile group action of pile 群桩效应
pile group interaction 群桩相互作用
pile guide 导桩
pile hammer 桩锤
pile head 桩头
pile heave 桩的隆起
pile hoop 桩箍
pile integrity test 桩的完整性试验
pile jacket 静力压桩机
pile jacketing 桩的套护
pile jacking method 桩压入法
pile jetting 射水冲桩

pile layout(plan) 桩位布置
pile loading test 桩静载试验
pile locking force 夹桩力
pile manufacture 桩的制作
pile penetration 桩的贯入
pile pier 桩式桥墩
pile point(tip, toe) 桩尖
pile pressing force measuring device 压桩力测量装置
pile pressing force transmitting system 压桩力传递系统
pile pulling 拔桩
pile retapping test 桩复打试验
pile shoe 桩靴
pile sinking method on the water 水上沉桩法
pile slab method 板桩加固法
pile spacing 桩距
pile splice 桩的拼接
pile stoppage point 桩止点
pile supported footing 桩承基础
pile supported raft 桩承筏基
pile test 桩基试验
pile tip 桩尖，桩端
pile with man-excavated shaft 人工挖孔灌注桩
pile with wings 带翼桩
pile work 打桩工程
piled wharf 桩式码头
pile-driving method 打桩方法
piles in row 排桩
pile-soil interaction 桩-土相互作用
pile-soil loading ratio 桩-土荷载比
pile-soil stress ratio 桩-土应力比
piling 打桩，打桩工程
piling method 打桩方法

piling noise 打桩噪声
piling rig 打桩机
piling underpinning 桩式托换
piling wall 桩墙
pilot pile 导桩
pilot test 中间试验
pilot tunnel method 导坑法，导洞法
pinch 捏，夹
pinhole test 针孔试验
pipe casing 管套，桩套管
pipe culvert 管涵
pipe drainage 管式排水
pipe jacking method 顶管法
pipe pile 管桩
pipe sinking 沉管法
pipe sleeve 管套
pipette method 移液管法
piping 管涌
piping erosion 管涌侵蚀
piping ratio 管涌比
pise 捣实黏土
pisolite 豆石
piston corer(sampler) 活塞取芯器
pit 坑，槽
pit exploitation 坑槽探，掘探
pit permeability test 试坑渗透试验
pit sampling 探槽取样
pit underpinning 坑式托换
pitching of pile 插桩，桩的斜度
pit-run gravel 天然级配砾石
placeability 可灌筑性，和易性
placement in layers 分层填筑
placement moisture 填筑湿度
placement water content 填筑含水量

placing under water 水下浇筑
plain concrete 素混凝土
plain fill 素填土
plain soil cushion 素土垫层
planar water 吸附水
plane deformation 平面变形
plane failure 平面破坏
plane failure analysis of rock slope 岩坡平面破坏分析
plane of break(rupture) 断裂面，破裂面
plane of equal settlement 等沉面
plane of saturation 饱和面
plane of shear 剪切面
plane of sliding 滑动面
plane of symmetry 对称面
plane of tensile fracture 张裂面
plane of weakness 软弱面
plane section assumption 平截面假定
plane slide 平面滑动
plane strain 平面应变
plane strain apparatus 平面应变仪
plane strain compression test 平面应变压缩试验
plane strain extension test 平面应变拉伸试验
plane strain state 平面应变状态
plane strain test 平面应变试验
plane stress 平面应力
plane wave 平面波
planimeter 面积仪
plank pile 厚木板桩
plastic analysis method 塑性分析法
plastic band-shaped drain 塑料带排水
plastic board drain method 塑料板排水法
plastic clay 塑性黏土
plastic collapse-basic theorem 塑性极限分析定理

plastic constitutive relations 塑性本构关系
plastic deformation 塑性变形
plastic drain 塑料板排水
plastic drainage pipe 塑料排水管
plastic equilibrium 塑性平衡
plastic factor of section 截面塑性系数
plastic failure 塑性破坏
plastic filter cloth 塑料滤布
plastic flow 塑流
plastic flow curve 塑性流动曲线
plastic flow law 塑性流动法则
plastic hinge 塑性铰
plastic index of clay （黏土的）塑性指数
plastic instability 塑性失稳
plastic limit 塑限
plastic limit analysis 塑性极限分析
plastic limit bending moment 塑性极限弯矩
plastic limit load 塑性极限荷载
plastic limit state 塑性极限状态
plastic limit test 塑限试验
plastic membrane 塑料膜
plastic modulus 塑性模量
plastic potential 塑性势
plastic range 塑性范围
plastic rupture of rock 岩石塑性破坏
plastic soil 塑性土
plastic stage 屈服阶段
plastic state 塑性状态
plastic strain 塑性应变
plastic strength 塑性强度
plastic theory of failure 塑性破坏理论
plastic volumetric strain 塑性体积应变
plastic zone 塑性区
plastic-elastic body 弹塑性体

plastic-elastic deformation 弹塑性变形
plasticity 塑性
plasticity chart 塑性图
plasticity index 塑性指数
plasticity needle 塑性试针
plasticity of rock 岩石的塑性
plasticity of soil 土的塑性
plasticity stability 塑性稳定性
plasticizer 增塑剂
plastic-viscous flow 塑性-黏滞流动
plastoelasticity 弹塑性
plastometer 塑性计，塑度计
plate loading test 平板荷载试验
plate shale 板页岩
plate stone 板岩
plate theory 薄板理论
plateau 高原
platelet 片晶
platform of pile frame 桩架平台
platform of piles 桩承台
platform pile 平台桩
Pleistocene epoch 更新世，洪积世
pleistoseismic zone 强震带
pliability 柔韧性，可挠性
plug effect 土塞作用(打桩)
plugging 钻孔回填
plumb line 铅垂线
pluton 深成岩体
pluvial 洪积层
pluvial alluvial region 冲洪积扇
ply adhesion 层间附着力
pneumatic caisson 气压沉箱
pneumatic caisson foundation 沉箱基础
pneumatic drill 风钻

pneumatic hammer　气锤
pneumatic mortar　压力喷浆
pneumatic pick　松土机，风镐
pneumatic piezometer　气压式孔隙水压力仪
pneumatic pile　气压桩
pneumatic rock drill　风动凿岩机
pneumatic settlement cell　气压式沉降仪
pneumatic tyred roller　轮胎压路机
pneumatically applied concrete　喷射混凝土
pocket penetrometer　袖珍贯入仪
pocket shear meter　小型十字板仪
point bearing pile　端承桩，支承桩
point load　集中荷载，点荷载
point loading test　点荷载试验
point of action　作用点
point of change of gradient　坡度变点
point of pile　桩端
point of yielding　屈服点
point resistance force　总桩端阻力
point resistance pressure　单位端阻力
point-bearing capacity　桩端承力
pointed pile　尖头桩
poise, P　泊(黏度单位)
Poisson's ratio　泊松比
Poisson's ratio of rock　岩石的泊松比
Poisson's ratio of soil　土的泊松比
polar force　极力
polar molecule　极性分子
polar moment of inertia　极惯性矩
polarizability　极化性
polarized shear wave velocity　极化剪切波速
polarizing microscope　偏光显微镜
pollution of ground water　地下水污染
polydrill　多用钻孔

polymer stabilization 聚合物加固
Poncelet graphical construction 庞塞莱特图解法
ponding test 浸水试验
pontoon pile driving plant 水上打桩机
poor graded 级配不良的
poor graded aggregate 级配不良的集料
poor subgrade(subsoil) 软弱地基
poorly-graded soil 不良级配土
pore(water)pressure dissipation
　　　孔隙(水)压力消散
pore air pressure 孔隙气压力
pore fluid 孔隙流体
pore medium 孔隙介质
pore phase 孔隙相
pore pressure 孔隙压力
pore pressure coefficient(parameter)
　　　孔隙水压力系数
pore pressure measurement 孔隙压力量测
pore pressure ratio 孔隙压力比
pore ratio of soil 土的孔隙比
pore ratio-pressure curve 孔隙比与压力曲线
pore size distribution analysis 孔径分布分析
pore space 孔隙，孔隙空间
pore structure 孔隙结构
pore suction 孔隙吸力
pore water 孔隙水
pore water head 孔隙水头
pore water pressure 孔隙水压力
pore water pressure cell 孔隙水压力仪
pore water pressure dissipation 孔隙水压力消散
pore water tension 孔隙水张力
pore-water pressure gauge 孔隙水压力计
porosimeter 孔隙计
porosity 孔隙率

porous aggregate 多孔集料
porous foundation 多孔性地基
porous medium 多孔介质
porous rock 多孔岩石，透水岩石
porous soil 多孔土
porous stone 透水石
porous structure 多孔结构
porphyrite 玢岩，微闪长石
porphyry 斑岩
portable compacter 手提击实仪
portable dipmeter 携带式地下水位测定仪
portable shear apparatus 轻便剪切仪
portal of tunnel 隧洞入口
post earthquake 余震
post-buckling equilibrium state 后屈曲平衡状态
post-buckling strength 后屈曲强度
post-construction monitoring 工后监测
post-failure modulus of rock 岩石破裂后期模量
post-failure stress-strain curve of rock
　　　岩石应力-应变后期曲线
potamogenic deposit 河口沉积
potential drop 势降
potential energy 势能
potential energy of loads 荷载势能
potential failure surface 潜在破坏面
potential function 势函数
potential infiltration rate 潜在渗入强度
potential logging 电位测井
potential surface of sliding 潜在滑动面
potential swell 膨胀势
potential vertical rise(PVR) 竖胀潜量
potential volume change(PVC) 体变潜量
potentiometer surface 测压管水位
pothole 凹坑，孔洞

pounder 夯锤
pour concrete 浇灌混凝土
poured-in-place concrete 现场浇筑混凝土
pouring 浇灌，浇筑
power method 幂法
power rammer 动力夯
power shovel 机械铲，动力铲
power-driven Luoyang spoon 机动洛阳铲
pozzolan cement 火山灰水泥
p-q curve p-q 曲线
practical applications 实际应用
Prager model 普拉格模型
Prandtl bearing capacity theory 普朗特承载力理论
Prandtl plastic equilibrium theory
 普朗特塑性平衡理论
pre-applied force 预加力
preboring 预钻孔
pre-buckling equilibrium state 前屈曲平衡状态
precast concrete pile 预制混凝土桩
precast concrete sheet pile 预制混凝土板桩
precast reinforced concrete pile
 预制钢筋混凝土桩
precast reinforced concrete pipe-pile
 预制钢筋混凝土管桩
precast reinforced concrete solid pile
 预制钢筋混凝土实心桩
precast reinforced concrete square-pile
 预制钢筋混凝土方桩
precautionary underpinning 预防性托换
precipitation 沉淀，降水量
precompression 预压
preconsolidated soil 先期固结土
preconsolidation 先期固结
preconsolidation pressure 先期固结压力

preconstruction fill 预压填土
predicted settlement 预计沉降量
predicted value 预测值
predominant period 卓越周期
predraining method 预先抽水法
predrilling 预钻孔
pre-excavation 预先开挖
prefabricated pile 预制桩
prefabricated strip drain 塑料排水带法
prefabricated subaqueous tunnel 预制管水底隧道
pre-formed pile 预制桩
preglacial deposit 前冰期沉积物
pre-hardening 初凝,初硬化
preliminary consolidation 初期固结
preliminary discrimination of liquefaction
  液化初步判别
preliminary grouting 预注浆
preliminary investigation 初步勘察
preliminary prospecting 初步勘探
preliminary shock 首震
preliminary symptom 地震前兆
preliminary test pile 预试桩
preliminary tremor 初期微震
preloaded soil 超固结土
preloading 堆载预压
preloading method 预压法
prepakt concrete 预填骨料混凝土,压浆混凝土
preservative treatment 防腐处理
preshear 预剪
Pre-Sinian period 前震旦纪
presplit blasting 预裂爆破法
presplitting 预裂法(岩石开挖的方法)
pressed pile 压入桩
pressed sampler 压入式取样器

pressiometer 压力仪,旁压仪
pressure arch 压力拱
pressure bulb 压力泡
pressure cell 压力盒
pressure coefficient 压力系数
pressure cushion 压力枕
pressure distribution 压力分布
pressure distribution angle of masonry foundation
　　(砌体基础的)刚性角
pressure drop 压力下降
pressure flow 有压流
pressure gage 测压计
pressure gauge
　　压力计,压力表,测压表,增压表
pressure grout pipe 压力灌浆管
pressure grouting 压力灌浆法
pressure head 压强水头
pressure hole 测压孔
pressure injection 压力灌注(法)
pressure measurement probe 测压管
pressure membrane apparatus 负压测定仪
pressure of surrounding rock 围岩压力
pressure of water flow 流水压力
pressure on foundation soil 基底压力
pressure pile 压力灌注桩
pressure relief well 减压井
pressure ring 压力环
pressure sensor 压力传感器
pressure tapping hole 测压孔
pressure test 压力试验
pressure transducer 压力传感器
pressure wave 压力波
pressuremeter 旁压仪,压力表
pressuremeter limit pressure 旁压仪极限压力

pressuremeter modulus　旁压仪模量
pressuremeter test(PMT)　旁压试验
pressure-void ratio curve　压缩曲线
prestressed anchor pile-slab wall　锚拉式板桩墙
prestressed concrete drilled caisson
　　　预应力混凝土管柱
prestressed concrete pile　预应力混凝土桩
prestressed ground anchor bar　预应力锚杆
prestressed loss　预应力损失
prestressed rock anchors　预应力岩石锚杆
prestressed soil anchor　预应力土锚
prestressed-concrete cylinder　预应力混凝土管桩
prestressing loss　预应力损失
prestressing method　预加应力法
presumptive bearing capacity　假设承载力
pretest pile　预试桩(托换工程)
prewetting　预浸水
primary clay　原生黏土
primary concretion　原生结核
primary consolidation　主固结
primary creep　主蠕变
primary creep of rock　岩石初始蠕变
primary earthquake　初震
primary liquidation　初始液化
primary loess　原生黄土
primary mineral　原生矿物
primary sedimentary structure　原始沉积构造
primary stress　初始应力(地应力)
primary time effect　主时间效应
primary valence bond　主价键
primary wave　初波，P波
principal axes of inertia　惯性主轴
principal axes of strain　主应变轴
principal axes of stress　主应力轴

principal compressive stress 主压应力
principal earthquake 主震
principal of continuity 连续性原理
principal of effective stress 有效应力原理
principal of equivalent loads 等效荷载原理
principal of minimum complementary energy
　　　最小余能原理
principal of minimum potential energy
　　　最小势能原理
principal of similitude 相似理论
principal of stationary potential energy
　　　势能驻值原理
principal of superposition 叠加原理
principal of virtual work 虚功原理
principal plane 主平面
principal shearing stress 主剪应力
principal strain 主应变
principal stress 主应力
principal stress circle 主应力圆
principal stress method 主应力法
principal stress ratio 主应力比
principal stress space 主应力空间
prismatic compass 棱镜罗盘
probabilistic analysis 概率分析
probabilistic design 概率设计
probabilistic uncertainty 概率不定性
probe 探头，探针
probing 探测，钻探
problem of plane strain 平面应变问题
problem of plane stress 平面应力问题
process of consolidation 固结过程
processed bentonite 加工膨润土
processed rock 加工石料
Proctor compaction curve 普罗克特击实曲线

Proctor compaction test	普罗克特击实试验
Proctor needle moisture test	普罗克特针测含水量试验
profile	剖面图
profilometer	粗糙度测定计
prognosis	预测
programming	程序设计
progressive failure	渐进破坏
progressive settlement	渐进性沉降，累计性沉降
progressive sliding	渐进性滑动
project	工程，计划，设计
projectile penetration	发射触探
proluvium	洪积层
proof load	检验荷载（桩基）
proof rest	验证性试验
propagation of error	误差传播
proportional limit	比例极限
proportional loading	比例加载
propping	支撑
prospect hole	勘探孔
prospect pit	探坑
prospecting	勘探
prospecting hammer	地质锤
prospecting shaft	勘探竖井，探井
protection of shallow foundation	浅基防护
protective filter	反滤层
Proterozoic era	元古代
Protodyakonov number	普氏系数，岩石强度系数
Protodyakonoves' number	普罗托吉雅可诺夫数
Protodyakonov's theory	普氏压力拱理论
proton magnetometer	质子磁力仪
prototype drill	原型钻头
prototype model	原型模型，足尺模型
prototype monitoring	原型监测

prototype observation 原型观测
prototype test 原型试验
proving ring 量力环，测力环
psephicity 磨圆度
psephite 碎砾岩，砾质岩
psephyte 砾质岩，砾状岩
pseudo-cohesion 准黏聚力
pseudo-consolidation pressure 准固结压力
pseudo-plastic fluid 拟塑性流体
pseudo-static approach 准静力法
pseudo-strength of rock mass 准岩体强度
pseudo-three-dimensional consolidation theory
    准三向固结理论
psychrometer 湿度计
puddle clay 捣实黏土
pull out force 拔力
pull strength 抗拔强度
pulling resistance 抗拔力
pulling test 张拉试验
pulling test on pile 桩的抗拔试验
pull-out load 拔出荷载
pull-out test 抗拔试验
pulp 泥浆
pulsating screen 振动筛
pulse generator 脉冲发生器
pulse load 脉冲荷载
pulse stress 脉冲应力
pulvimixer 松土搅拌机
pumice 浮石，轻石
pumice soil 轻石土
pump-in test 压水试验
pumping depression well 抽水减压井
pumping from well point 井点抽水
pumping test 抽水试验

punch capacity 冲切承载力
punch shear test 冲剪试验
punching hole 冲孔
punching of pile on cap 桩对承台的冲切
punching shear 冲剪
punching shear area 冲切面积
punching shear failure 冲剪破坏,地基刺入剪切破坏
punching shear stress 冲剪应力
puncture resistance 击穿阻力
puncture strength 抗刺强度
pure shear 纯剪
pushover analysis 静力弹塑性分析
putrid mud 腐泥
P-wave P波,压缩波
P-Y curve method P-Y 曲线法
pycnometer 比重瓶
pyrite 黄铁矿
pyroxene 辉石

Q

Q(quick direct shear) test 快剪试验
QM anchorage QM 型锚具
qualification test 鉴定试验
quality control 质量控制
quality control of earth-rock fill 填筑质量控制
quantity of percolation 渗流量
quarry 采石场
quarry stone 毛石
quartering 四分法
quartz 石英
quartz sand 石英砂

quartzite 石英岩
quasi-overconsolidation 准超固结
quasi-permanent combination 准永久组合
quasi-permanent value 准永久值
quasi-preconsolidation pressure 准先期固结压力
quasi-saturated soil 准饱和土
quasi-static cone penetration 准静力触探贯入度
quasi-static method 准静力法
quaternary 第四纪
Quaternary alluvium 第四季冲积层
quaternary deposit 第四纪沉积物
quaternary geological map 第四纪地质图
quaternary geology 第四纪土质学
quaternary glacier 第四纪冰川
Quer-wave 奎尔波
quick clay 过敏性黏土
quick consolidation test 快速固结试验
quick sand 流砂
quick shear test 快剪试验
quick soil classification 土的简易分类法
quick test 快速试验,快剪试验
quicklime 生石灰
quicklime pile 石灰桩
quincuncial piles 梅花桩
Q-wave Q波,乐普波

R

R_0-test R_0-试验
rabbling mechanism 搅拌机械
racking pile 斜桩
radial bracing 径向支撑
radial compressive stress 径向压应力

radial consolidation 径向固结
radial crack 径向裂纹
radial crushing strength 径向抗压强度
radial deformation 径向变形
radial displacement 径向位移
radial distribution function 径向分布函数
radial drainage 径向排水
radial fissure 径向裂缝
radial flat jack technique 径向扁千斤顶法
radial flow 径向流动
radial force 径向力
radial normal stress 径向正应力
radial principal stress 径向主应力
radial section 径切面
radial shear 径向剪切
radial strain 径向应变
radial stress 径向应力
radial support 径向支撑
radial tension 径向张力
radial vector 径向矢量
radial velocity 径向速度
radial wells 辐射井
radiation wall 防辐射墙
radial-thrust force 径向抗力
radius deformation 回弹变形
radius of influence 影响半径
radius of grouting spread 浆液扩散半径
radius of plastic zone 塑性区半径
radius of seismic danger due to blasting
　　　　　爆破地震危险半径
radius strain 回弹应变
radius stress 回弹应力
radius vector 矢径
raft foundation 筏形基础

raft foundation under walls 墙下筏形基础
rafter set 基坑支架
rafter timbering 基坑支撑
rail beam 轨束梁
rail bearing 轨枕,轨承,轨底支承面
rail bed 钢轨底座
rail facility 铁路
railroad bed 铁路路基
railroad tunnel 铁路隧道
rails tie plate 垫板
railway bridge 铁路桥
railway crossing 铁路道口
railway culvert 铁路涵洞
rain canopy 防雨篷
ram 夯锤,锤体
ram engine 打桩机
ram guide 桩架
ram impact machine 捣锤冲击机,打夯机
ram machine 打桩机
ram steam pile driver 汽锤打桩机
ram stroke 冲程
rammed bulb pile 无桩靴夯扩灌注桩
rammed clay 夯实黏土
rammed concrete 夯实混凝土
rammed earth building 土建筑
rammed earth wall 夯土墙
rammed lime earth 灰土夯实
rammed-cement-soil pile 夯实水泥土桩法
rammed-earth 素土夯实
rammed-soil pile 夯实土桩
rammer compactor 夯土机,夯实机
ramming 打夯
ramming by heavy hammer 重锤夯实
ramming depth 打夯深度

ramming machine 捣固机，打夯机
ramming speed 抛砂速度，打夯速度，捣实速度
ramp slope 斜坡坡度
randanite 硅藻土
random arrangement 随机排列
random bulk 料堆
random coring 混凝土钻探抽样测试
random crack 不规则裂缝
random load 随机荷载
random masonry 乱砌圬工
random rubble 乱切毛石
random rubble wall 乱石墙
Rankine state 朗肯状态
Rankine's earth pressure theory
　　　朗肯（朗金）土压力理论
Rapakivi granite 环斑花岗石
rapid ascent 陡坡
rapid drawdown 水位骤降
rapid impact rammer 快速冲击夯实
rapid setting admixture 速凝剂
rapid soil classification 土的简易分类法
rate of consolidation 固结速率
rate of crack propagation 裂缝扩展速率
rate of creep 蠕变率
rate of diffusion 扩散速率
rate of expansion 膨胀率
rate of flow 流速
rate of free expansion 自由膨胀率
rate of grout acceptance 吸浆量，吃浆量
rate of penetration 贯入速率
rate of percolation 渗透速率
rate of secondary consolidation 次固结速率
rate of settlement 沉降速率
rate of shear strain 剪应变速率

rate of specific surface area 比表面积率
rate of stripping 剥离比
rate of swelling of soil 土的膨胀率
ratio of elastic modulus 弹性模量比
ratio of length-diameter 长径比
ratio of pricipal stress 主应力比
ratio of rise to span 高跨比
ratio of secant modulus 割线模量比
ratio of slope 坡度比
Rayleigh wave 瑞利波，R 波
Raymond concrete pile 雷蒙式桩
reaction force 反作用力
real fluid 实际流体，黏性流体
real specific gravity 真密度(比重)
ream 锥孔(扩大)
reamer 扩孔机
reamer bit 扩孔钻头
rear arch 背拱
rear support 后支撑
rear-end 后端
rebound 回弹
rebound apparatus 回弹仪
rebound curve 回弹曲线
rebound deformation 回弹变形
rebound index 回弹指数
rebound method 回弹法
rebound modulus 回弹模量
rebound of foundation 地基回弹
rebound of pile 桩的回弹
rebound pendulum machine 回跳打桩机
rebound strain 弹性应变
receiving stream 排污河流
recently deposit soil 新近堆积土
recharge area 补给区

recharge method　回灌法
recharge rate　补给率
recharge well　回灌井
recharging　回灌
reciprocating activity　往返活动性
recirculation of air　循环风
reclaimed land　填海土地
reclamation　围垦工程
reclamation level　填筑高程
reclamation works　填海工程
recoil　反冲力
recompaction　再压缩
recompression curve　再压缩曲线
reconnaissance　勘测，踏勘
reconnaissance exploration　可行性研究勘查
reconnaissance phase　踏勘阶段
reconnaissance survey　查勘测量，路线测绘
reconsolidation　再固结
reconsolidation volumetric strain　再固结体应变
recorder well　观测井
recoverable deformation　弹性变形
recovery of elasticity　弹性变形
recrystallization　重结晶作用
rectangular cross section beam　矩形截面梁
rectangular element　矩形单元
rectangular foundation　矩形基础，矩形地基
rectangular girder　矩形梁
rectangular shield　矩形盾构
red clay　红黏土
red ochre grout　红土浆
redeposited loess　次生黄土
redeposition　再沉积
redistribution　再分配
redistribution of moment　弯矩再分配

redriving 复打(桩工的)
reduce section 变截面
reduced density 折算密度
reduced depth 折算深度
reduced elastic modulus 折算弹性模量
reduced factor 折减系数
reduced head 折算水头
reduced height 折算高度
reduced load 折算荷载
reduced stress 折算应力
reduced value 折算值
reduction factor for pile in group 群桩折减系数
reduction of an arbitrary force system 任意力系的简化
reduction of area 截面收缩率
reduction of area at fracture 断裂收缩率
reduction of erection time 工期缩短
refilling 回填
reflection survey 反射法勘探
refusal criterion 拒浆标准
refusal flow rate 拒浆流量
refusal pressure 拒浆压力
regional engineering geology 区域工程地质
regional hydrogeology 区域水文地质学
regional hydrology 区域水文学
regional settlement 区域性沉降
regional soil 地区性土
regolith 风化层
regrade 重整坡度,重整地形
regrouting 二次注浆
regulated set cement 调凝水泥
reinforced(earth)retaining wall 加筋土挡土墙
reinforced concrete arch culvert 钢筋混凝土拱涵
reinforced concrete base 钢筋混凝土地基

reinforced concrete caisson 钢筋混凝土沉箱
reinforced concrete column 钢筋混凝土柱
reinforced concrete open caisson
　　钢筋混凝土开口沉箱
reinforced concrete pile 钢筋混凝土桩
reinforced concrete pneumatic caisson
　　钢筋混凝土气压沉箱
reinforced concrete retaining wall
　　钢筋混凝土挡土墙
reinforced concrete shear wall
　　钢筋混凝土剪力墙
reinforced concrete sheet pile 钢筋混凝土板桩
reinforced concrete strip foundation under columns
　　柱下钢筋混凝土条形基础
reinforced earth 加筋土
reinforced earth abutment 加筋土桥台
reinforced embankment 加筋路堤
reinforced fill wall 加筋挡土墙
reinforced layer 加固层
reinforced rib 加强筋
reinforced soil wall 加筋土挡墙
reinforced stress 增强应力
reinforced structure 加固结构
reinforced stulls 斜撑加强支架
reinforcement for shearing 抗剪加固
reinforcement method 加筋法
reinforcement of soft foundation 软基加筋
reinforcement plate 加固板，增强板
reinforcement ratio 配筋率
reinforcement ratio per unit volume 体积配筋率
reinforcing dam 加固堤
reinforcing filler 增强充填
reinforcing girder 加强梁
reinforcing plate 加强板

reinforcing rib 加强肋
reinforcing stay 加固支撑
relationship between water and sediment 水沙关系
relative compaction 相对密实度
relative consistency 稠度指数
relative density 相对密度
relative density of soil 土的相对密度
relative density test 相对密度试验
relative depth of compressive area
　　　　　相对受压区高度
relative displacement 相对位移
relative elevation 相对高程
relative embedment 相对埋深
relative error 相对误差
relative expansion 相对膨胀
relative hardness 相对硬度
relative humidity 相对湿度
relative ice content 相对含冰量
relative plasticity index 稠度指数
relative settlement 相对沉降量
relative surface area 比表面
relative water content 液性指数
relaxation effect 松弛效应，张弛效应
relaxation factor 松弛因数，松弛因子
relaxation function 松弛函数
relaxation loss 张弛损失
relaxation modulus 松弛模量
relaxation of rock 岩石应力松弛
relaxation of stress 应力松弛
relaxation test 松弛试验
relaxation time 松弛时间
relaxed rock 松弛岩体
release joint 释放节理
release of stress 消除应力

release point 出逸点
reliability coefficient 可靠性系数
reliability design 可靠性设计
reliability index 可靠指标
reliability of structure 结构的可靠性
reliability trials 可靠性试验
reliability-based optimum design
   基于可靠度的优化设计
relic structure 残余结构
relict bedding 残留层理
relief platform 卸荷台
relief well 减压井
reloading 再加载
remaining rock 残岩
remaining stress 残余应力
remedial grouting 补救灌浆，补强灌浆
remedial treatment 补强处理
remedy 补救，修补
remoldability 重塑性
remolded clay 重塑黏土
remolded properties 重塑土特性
remolded soil 重塑土
remolded strength 重塑强度
remolding 重塑，扰动
remote sensing prospecting 遥感勘测
renewed fault 复活断层
repeated bending fatigue 反复弯曲疲劳
repeated direct shear test 反复直剪强度试验
repeated loading 重复荷载
repeated stress fatigue 疲劳断裂
repeated-load test 重复试验
replaced foundation 置换基础
replaced ground base 置换地基
replacement method 置换法

replacement of foundation 基础置换
replacement pile 置换柱
replacement ratio 置换率(复合地基)
report of engineering geologic investigation
　　　工程地质测绘勘查报告
reprecipitation 再沉积
representative value of load 荷载代表值
resaturation 再饱和
resedimented rock 再沉积岩
reserved strength 保留强度
residual analysis 残差分析
residual angle of internal friction 残余内摩擦角
residual clay 原生土
residual cohesion intercept 残余黏聚力
residual deflection 残余挠度
residual deformation 残余变形
residual deformation of soil mass 土体的残余变形
residual deposite 残余沉积，残积物
residual diluvial expansive soil 残坡积膨胀土
residual moisture 残余水分
residual pore water pressure 残余孔隙水压力
residual saturation 残余饱和度
residual settlement 残余沉降
residual shear strength 残余抗剪强度
residual shrinkage 残余收缩
residual soil 残积土
residual strain 残余应变
residual strain field 残余应变场
residual strength 残余强度
residual stress 残余应力
residual stress field 残余应力场
residual stress in rock 岩石残余应力
residual tensile strength 残余抗拉强度
residual value 残值

residue of boundary 边界余量
residue of equation 方程余量
resilience 回弹
resilience index 回弹指数
resiliometer 回弹仪
resin bolting 树脂锚杆
resistance bond 抗粘性
resistance factor 阻抗系数
resistance of piles 群桩抗力
resistance to acid 耐酸性
resistance to alternate freezing and thawing 抗冻融性
resistance to alternate mechanical stress 耐交变应力性
resistance to deformation 抵抗变形
resistance to failure 抗破坏能力
resistance to impact 抗冲击强度
resistance to lateral bending 抗侧弯能力
resistance to sliding 抗滑稳定性
resistance to splitting 抗劈力
resistance to water 耐水性
resistance to water vapor permeability 蒸汽渗透阻
resistance(resistant force) 抗力
resistant structure 抗力结构
resistant zone 抗拔区(加筋土)
resisting moment arm method 内力臂法
resisting power 抵抗力
resistivity 电阻率
resistivity curve 阻力曲线
resistivity exploration 电阻法
resistivity log 电法测井
resistivity method 电阻率法
resolution of a force 力的分解
resonance method 共振法

resonant column test 共振柱试验
rest water level 井内静水位
restoring force 恢复力
restoring force character 恢复力特性
restoring model 恢复力模型
restrain 约束
restrain condition of pile top 桩顶约束条件
restrain end 约束端点
restrain joint 约束节点
restrained end 固定端
restrained pile 约束桩
restrained torsion 约束扭转
restrained torsional constant 约束扭转常数
restraining moment 约束力矩
restraint coefficient 约束系数
restraint condition 约束条件
restraint deformation 约束变形
restraint stress 约束应力
resultant(resulting force) 合力
resulting force curve 合力曲线
resurfacing 表面处理
retained material on the sieve 筛余物
retained percentage 筛余百分率
retained water 吸附水
retaining and protection of building foundation excavation 基坑支护
retaining dam 挡水坝
retaining pile 护壁桩
retaining slab 挡土板
retaining structure 挡土构筑物
retaining wall 挡土墙,挡墙
retardation 滞后
reticular structure 网状结构
reticulated masonry 网式砌墙

reticulated root piles 网络树根桩
retraction stress 收缩应力
retreatment 再处理
retrogression of strength 强度退化
retrogressive slope 反坡
reverse bend 反向弯矩
reverse bend test 反向弯矩试验
reverse building method 逆作法
reverse compression 反向压力
reverse fault 逆断层
reverse work 反向工作法
reversible expansion 可逆膨胀
reversing shear box test 往复直剪试验
revetment 护坡,护墙
revetment bank 护岸
revetment dike 护堤
revetment wall 护墙,护岸墙
rheological behaviour(of mortar or slurry)
　　(水泥砂浆或泥浆的)流变性
rheological model(of rock or soil)
　　(岩石或土的)流变学模型
rheological pressure of surrounding rock
　　围岩流变压
rheology 流变学
rhyolite 流纹岩
rhyolitic lava flow 流纹岩的熔岩流
ribbon clay 带状黏土
rice-straw ash 稻草灰
rice-straw board 稻草板
rich clay 富黏土
rickle 松散堆积物
rigid base 刚性地基
rigid connection 刚性连接
rigid core wall 刚性心墙

rigid culvert   刚性涵洞
rigid fill   刚性垫层
rigid footing(foundation)   刚性基础
rigid guide-rod   刚性导向钻杆
rigid joint between tube segments   管段刚性接头
rigid pier   刚性墩
rigid plastic model   刚塑性模型
rigidity modulus   刚性模量
rim angle   边界角,接触角
ring(shaped)foundation   环形基础
ring shear test   环剪试验
ripper equipment   松土器
riprap   乱石护坡
riprap protection   抛石护岸
riprap protection of slope   抛石护坡
riprap stone revetment   堆石护坡
riprap work   抛石工程
rise of arch   拱矢
riser pipe   井点管
rising shaft   竖井
river basin   河流流域
river bed   河床
river drift   河流沉积
river dynamics   河流动力学
river engineering   河流工程学
river erosion   河流侵蚀
river fan   河流冲积扇
river sediment   河流泥沙
road bed(roadbed)   路基
road pavement   道路路面
roadbed section   路基断面
road-metal   路碴,筑路碎石
rock   岩石
rock abrasion platform   浪蚀岩石阶地

rock analysis 岩石分析
rock anchor 石锚，入石锚定杆
rock and soil 岩土
rock and soil engineering 工程岩土学
rock and soil mechanical parameters 岩土力学参数
rock avalanche 岩崩
rock bar 岩坝
rock bedding 岩层
rock blast 凿岩爆破
rock block 块石
rock block toppling 岩块式倾倒
rock block with preconditioned smooth surface 光面岩石
rock block with rough and multiple 毛面岩石
rock bolt 岩石锚拴
rock bolt dynamometer 岩石锚杆测力计
rock bolt extensometer 岩石锚杆伸长仪
rock bolter 锚杆冲击机
rock bolting 岩石锚杆
rock bolting theory 岩石锚杆理论
rock bonding 岩石固结
rock boring machine 凿岩机
rock bump 岩石突出
rock burst 岩爆
rock classification 岩石分类
rock cleavage 岩石劈理
rock core 岩芯
rock core bit 取芯钻头
rock cover 岩石覆盖层
rock creep 岩石蠕变
rock crushing strength 岩石破碎强度
rock cut slope 削石坡，岩削坡
rock damage 岩体损伤
rock decay 岩石风化

rock discontinuities 岩体不连续面
rock dowel 石钉，岩石销钉
rock drain 填石盲沟
rock drill mounting 钻架
rock element 岩石要素
rock engineering 岩石工程
rock fabric 岩组学
rock facing 堆石护坡
rock fall 岩崩
rock fan 石质扇形地
rock fissure 岩石裂隙
rock formation 岩层
rock forming mineral 岩石矿物
rock foundation 岩石地基
rock foundation grouting 岩基灌浆
rock fracture mechanics 岩石断裂力学
rock grouting 岩石灌浆
rock hound 地质学家
rock house 岩洞
rock in place 原生岩石
rock joint 岩石节理
rock joint shear strength test 岩石节理剪力试验
rock mass 岩体
rock mass basic quality(BQ) 岩体基本质量
rock mass dynamics 岩石动力学
rock mass factor 岩石系数
rock mass strength 岩体强度
rock mass stress test 岩体应力测试
rock matrix compressibility 岩石基质压缩系数
rock mechanics 岩石力学，岩体力学
rock mound 岩石墩，堆石
rock outcrop 岩石露头
rock penetration performance 凿岩效率
rock picker 风镐

rock pillar 岩柱
rock pin 岩石插筋锚固
rock plug blasting 岩塞爆破
rock pocket 岩穴
rock pressure 山岩压力
rock pressure in hill 山体压力
rock pressure in horizontal roadway 平巷地压
rock pressure in inclined shaft 斜井地压
rock pressure in vertical shaft 立井地压
rock pressure indications 岩石压力迹象
rock pressure theory 岩石压力理论
rock prestressing 岩石预应力
rock prestressing theory 岩石预应力假说
rock quality designation(RQD) 岩石质量指标
rock quality index 岩石质量系数
rock rating 岩石分级
rock roof bolting 岩顶锚杆支撑
rock slide 岩崩,岩滑,落石,坍方
rock slope 岩石坡,石坡
rock slope stability 岩石坡稳定性
rock stratification 岩层,岩石层理
rock stratum 岩层
rock strength 岩石强度
rock stress 岩石应力
rock structure 岩石构造
rock surface 岩面
rock testing 岩石试验
rock texture 岩石结构
rock type 岩石类型
rock weathering 岩石风化
rock yield 岩石屈服
rockbolt 锚杆
rock-bolt supporting 岩锚支护
rock-face 岩石立面

rockfill dam 堆石坝
rock-forming mineral 造岩矿物
rockhole 岩层钻孔
rocking stiffness factor 地基抗弯刚度系数
Rockwell apparatus 洛氏硬度计
Rockwell hardness number 洛氏硬度值
Rockwell hardness tester 洛氏硬度试验机
Rockwell indentation hardness 洛氏针入硬度
rod 钻杆
rod rotation speed 钻杆转速
rod sounding 钎探
rolled fill earth dam 碾压土坝
rolled-earth fill 碾压填土
roller compaction 碾压
roller compression 碾压法
rolling compaction test 碾压试验
rolling topography 起伏地形
rolling trial 碾压试验
roof clay 隔水黏土
roof control 顶板控制
root pile 树根桩
rose diagram of joints 节理玫瑰图
roseite 蛭石
rotary cutting test on the rock by the disc rolling cutter 盘形滚刀回转切割岩石试验
rotary inertial 转动惯量
rotation speed for drilling 钻进转速
rotatory boring cast-in-place pile 回转钻孔灌注桩
roughing stone 毛料石
roughness 粗糙度
roughness factor 粗糙系数
roughness of the surface 表面粗糙度
roundabout open caisson method 环形沉井法
round-hole sieve 圆孔筛

route reconnaissance 路线踏勘
routine analysis 常规分析
routine consolidation test 常规固结试验
routine monitoring 常规监测
routine observation 常规观测
row of piles 排桩
R-test 固结快剪试验
rubble 毛石
rubble aggregate 毛石集料
rubble concrete 块石混凝土
rubble foundation 块石基础
rubble retaining wall 毛石挡土墙
rubble stone masonry 毛石砌体
rubble wall 毛石墙
rubble work 毛石工程
rubble-mound foundation 抛石基床
running ground 流沙地基
running sand 流砂
runoff 径流
runoff area 径流区
runoff coefficient 径流系数
runoff computing formula 径流计算公式
runoff in depth 地下径流
rupture 断裂
rupture of shaft lining 井壁破裂
rupture plane of slope 土滑动面
rupture strength 破裂强度
rupture stress 破裂应力

S

saddle-shaped shell 鞍形壳
safe bearing capacity 安全承载能力

safe drilling 安全钻井
safe factor 安全系数
safe formation 安全地层
safe load 容许荷载
safe margin 安全储备
safe measure 安全措施
safety against overturning 抗倾覆安全
safety against shear failure 抗剪切破坏安全
safety against sliding 抗滑安全
safety analysis 安全分析
safety coefficient/factor 安全系数
safety devices of crane 起重机安全装置
safety distance 安全距离
safety factor against shear failure
　　　剪切破坏安全系数
safety factor for bearing capacity 承载力安全系数
safety factor of slope 边坡安全系数
safety factor of stability 稳定安全系数
safety factor of stability of slope
　　　边坡稳定安全系数
safety load carrying capacity 安全承载能力
sagging soil 融沉土
Saint-Venant's principle 圣维南原理
saline soil 盐渍土
saline-alkali soil 盐碱土
salinity 盐度，含盐量
salt content 含盐量，盐度
salt ground water 含盐地下水
salt-bearing rock 含盐类岩石
salt-type flocculation 盐型絮凝
salty soil 盐渍土
sample covariance 样本协方差
sample dispersion 试验分散性
sample divider 分土器

sample extrude 推样土器
sample of undisturbed soil 原状土试样
sample splitter 分样器
sampling 采样
sampling action 取样
sampling disturbance 取样扰动
sampling hole 测孔
sampling length 取样长度
sampling observation 取样观测
sampling size 样本大小
sampling tube 取样管
sampling well 取样井
sand 砂土
sand and gravel 砂砾
sand and gravel overlay 砂砾覆盖层
sand and gravel trap 沉砂池(井)
sand asphalt 沥青砂
sand avalanche 砂崩
sand bag 砂袋
sand bailer 抽砂筒
sand bedding course 砂垫层
sand bitumen 沥青砂
sand blanket 砂垫层
sand blasting 喷砂处理
sand boiling 砂沸
sand casting cement 砂型水泥
sand cement 掺砂水泥
sand cement ratio 砂灰比
sand chimney 砂质垂直排水
sand coarse aggregate ratio(S/A) 砂石率
sand column 砂桩
sand compaction pile 砂桩挤密
sand compaction pile method 挤密砂桩加固法
sand cone method 灌砂法

sand consolidation anchorage	砂固结锚固
sand content	砂粒含量
sand cushion	砂垫层
sand cushion stabilization method	砂垫层加固法
sand cutting	拌砂
sand drain	排水砂井
sand drain method	砂井排水法
sand drain vacuum method	砂井真空排水法
sand filter	砂滤池
sand filtration	砂滤法
sand flow method	压砂法
sand grading	砂粒分级
sand grouting	砂浆
sand heap analogy	砂堆比拟
sand mat	砂垫层
sand mortar	砂浆
sand pile	砂桩
sand pile stabilization method	砂桩加固法
sand piling barge	砂桩船
sand pump	砂泵
sand relative density test	砂的相对密实度试验
sand replacement method	换砂法
sand screen	筛砂
sand setting	沉砂
sand soil	砂土
sand wick	袋装砂井
sandbag revetment	砂袋护坡
sandbag walling	砂袋筑墙
sand-blasting	喷砂
sand-blasting method	喷砂处理法
sand-coarse aggregate ratio	砂石比
sand-filled drainage well	沙井，填沙排水井
sand-gravel pile	砂石桩
sandpad	砂垫层

sandstone	砂岩
sandwich structure	夹层结构
sandwich waterproofing of lining	衬砌夹层防水
sandy aggregate	砂质集料
sandy clay	砂质黏土
sandy gravel	砂砾石
sandy gravel layer	砂砾石垫层
sandy limestone	砂质石灰岩
sandy seabed	砂质海床
sandy shale	砂质页岩
sandy silt	砂质粉土，亚砂土
sandy soil	砂井土，砂类土
sandy soil foundation	砂土地基
saprolite	风化土
saprolith	残余土
saprolitic	腐泥土
saturated	饱和的
saturated belt	饱和带
saturated density	饱和密度
saturated liquid	饱和液
saturated permeability	饱和透水性
saturated sand	饱和砂
saturated soil	饱和土
saturated solution	饱和溶液
saturated state	饱和状态
saturated steam	饱和蒸汽
saturated surface	饱和面
saturated unit weight	饱和重度
saturated unit weight of soil	土的饱和容重
saturated water	饱和水
saturated zone	饱和层
saturated zone monitoring well	饱和区监测井
saturation	饱和
saturation analysis	饱和分析法

saturation capacity 饱和容重，饱和含水量
saturation coefficient 饱和系数
saturation concentration 饱和浓度
saturation condition 饱和状态
saturation curve 饱和曲线
saturation deficiency(deficit) 饱和差
saturation degree 饱和度
saturation density of soil 土的饱和密度
saturation field 饱和场
saturation force 饱和力
saturation gradient 饱和梯度
saturation humidity ratio 饱和含湿量
saturation index 饱和指数
saturation line 浸润线，饱和曲线(击实试验)
saturation moisture content 饱和含水率
saturation of soil 土的饱和度
saturation percentage 饱和百分率
saturation point 饱和点
saturation pressure 饱和压力
saturation ratio 饱和比
saturation specific humidity 饱和比湿度
saturation steam 饱和蒸汽
saturation steam temperature 饱和蒸汽温度
saturation swelling stress 饱和湿胀应力
saturation temperature 饱和温度
saturation test 饱和程度试验
saturation testing 饱和测试
saturation time 饱和时间
saturation unit weight of soil
　　　　　土的饱和容重(表观密度)
saturation weight density of soil 土的饱和重度
saturation zone 饱和区，饱和层
scale effect 尺度效应，尺寸效应，放缩效应
scale factor 相似系数，比例因子

scaling 定比例
scaling factor 缩比例
scarp wall 陡坡墙
scattering stone 飞石
schist 片岩
schistous sands 片状砂岩
scour 冲刷
scouring force 冲刷力
scraping blade 刮土板
scraping cutter 刮刀
scraping grader 刮土地平机
scraping hardness test 划痕测硬法
scraping pebble 擦痕卵石
scree 崩落
screed board 刮板
screen 筛选
screen analysis 颗粒分析
screen classification 筛分
screen of deep well 深井滤水管
screen size 筛分粒度
screen size gradation 筛分级配
screen sizing 筛分分级
screen tailings 筛余物
screen test 筛分试验
screen well point 滤网式井点
screened sand 过筛砂
screened well 过滤井
screening 筛分法
screening inspection 筛分检查
screw 螺钉
screw anchor 螺钉锚固,带螺纹锚桩
screw holder 钻杆导向器
screw pile 螺旋桩
screw plate test 螺旋板载荷试验

sea beach 海滩
seal coat 止水层,防水层
seal grouting 密封灌浆
seal installation 止水设施
seal with fines 细粒料填缝
sealed earth 封填土
sealed joint 止水缝,填实缝
sealed shoulder 加固的路肩
sealed up concrete curing 密闭状态养护
sealing anchorage at beam end 封锚
sealing clearance at the tail of shield 盾构密封间隙
sealing device at the tail of shield 盾尾密封装置
seal-level datum 高程基准面
season crack 干裂
seasonal depletion 季节性枯竭
seasonal load 季节性荷载
seasonal reservoir 季节调节水库
seasonal stream 季节性河流
seasonal variation factor 季节性变化系数
seasonally frozen soil 季节冻土
seat angle 座角钢
seat beam 座梁
seat board 座板
seat earth 底板,底层
seat stay 座撑
seawall 海堤
secant circle 正割圆
secant curve 正割曲线
secant method 正割刚度法
secant modulus 割线模量
secant modulus of elasticity 正割弹性模量
secant modulus of rock 岩石割线模量
secant pile wall 搭接桩墙
secant yield stress 正割屈服应力

secondary blasting 二次爆破
secondary cast 二次浇筑
secondary clay 次生黏土
secondary consolidation 次固结
secondary consolidation settlement 次固结沉降
secondary expansion 二次膨胀
secondary grouting 二次压注
secondary identification of liquefaction
 液化的再判别
secondary loading 附加荷载
secondary mineral 次生矿物
secondary moment 附加力矩
secondary principal stress 次主应力
secondary story in tunnel timbering 中层支撑
secondary stress 次应力
secondary structure 次生结构面
section 剖面，截面
section area of reinforcement 钢筋横断面
section diagram 横断面图
section modulus for a beam 抗弯截面模量
section moment 截面矩
section plastic modulus 塑性截面矩
section well 吸水井
sectional area 截面积
sectional construction 预制构件拼装结构
sectional dimension 截面尺寸
sectional drawing 截面图，断面图
sectional elevation 立面图
sectional retaining wall 分段挡土墙
sectional sizing variables 截面尺寸变量
sectional view 剖视图
sector distribution 扇形分布
sector coordinate 扇形坐标
sector coordinate system 扇形坐标系统

sector theory 扇形理论
sectorial geometric property 扇形几何性质
sectorial moment of inertia 扇形惯性矩
sectorial normal stress 扇形正应力
sectorial shear stress 扇形剪应力
sectorial statical moment of area 扇形静面矩
secular coldest month 累年最冷月
sedentary soil 原生土
sediment 沉积物
sediment analysis 沉淀法粒度分析
sediment concentration 含沙量
sediment concentration as percentage by weight 含沙重量比浓度
sediment deposition 泥沙沉降
sediment load 输沙量，河流泥沙
sedimentary rock 沉积岩
sedimentary soil 沉积土
sedimentary tectonics 沉积构造
sedimentation 沉淀
sedimentation curve 沉积曲线
seeded slope 植草护坡
seep proof screen 止水墙
seepage 渗流，渗出量
seepage area of well 井的渗流面积
seepage around abutment 绕坝渗漏
seepage coefficient 渗透系数
seepage control 渗流控制
seepage deformation 渗透变形
seepage discharge 渗流量
seepage failure 渗透破坏
seepage force 渗透力
seepage loss 渗透损失
seepage path 渗径
seepage pressure 渗透压力

seepage prevention 防渗
seepage proof curtain 防渗帷幕
seepage quantity 渗流量
seepage resistance of concrete 混凝土抗渗性
seepage velocity 渗流速度
seepage-flow model 渗流模型
seepage-off 渗出
seep-in grouting 渗入性灌浆
seepy material 透水材料
segment 部分，图段
segmental arch 弓形拱
segmental arch culvert 弓形拱涵
segregation 离析
segregation index 离析指数
segregation of concrete 混凝土离析
segregation of mortar 砂浆分层度
segregation zone 离析区
seismic acceleration 震动加速度
seismic action effect 震动效应
seismic analysis 震动分析
seismic appraise 抗震分析
seismic bedrock motion 基岩地震动
seismic exploration 地震勘探
seismic force 地震力
seismic predominant period 地震卓越周期
seismic prospecting 地震勘探
seismic subsidence 震陷
seismic survey 地震查勘
self weight collapse loess 自重湿陷性黄土
self weight non-collapse loess 非自重湿陷性黄土
self-boring pressure meter 自钻式旁压仪
self-cement for stressing 自应力水泥
self-consolidation 自重固结
self-stiffness 固有刚度

self-stressing method 自应力法
self-supporting static pressed pile 自承式静压桩
self-weight stress 自重应力
semi-empirical formula 半经验公式
semi-hard rock 半坚硬岩石
semi-infinite elastic body 半无限弹性体
semi-infinite plane problem 半无限平面问题
semi-infinite solid 半无限体
semi-inverse method 半逆解法
semi-mechanized shield 半机械化盾构
semi-saturation 半饱和
semi-shield 半盾构
sensibility 灵敏度
sensibility analysis 灵敏度分析
sensibility of shield 盾构灵敏度
sensitive clay 灵敏黏土
sensitive to heat 热敏的
sensitivity 灵敏度
sensitivity of soil 土的灵敏度
sensor 传感器
separation distance 安全距离
sequence of filling 填筑顺序
sequence of grouting 灌浆顺序
sequence of pours 浇筑顺序
serial construction 连续施工
serpentine boulder 蛇纹石
service life 使用年限
serviceability limit state 正常使用极限状态
serviceability stage 正常使用阶段
set grout 灌浆结石
set of pile 桩的贯入度
set-retarding 缓凝作用
set-retarding admixture 缓凝剂
setting 凝结

setting expansion 凝结膨胀
setting of plane-table 平板仪安装
setting out 施工放样
setting out of axis 轴线放样
setting out of building 建筑放线
setting out of foundation 基础放样
setting out rod 定线桩
setting stake 定线桩
setting strength 凝固强度
setting time 凝固时间
setting up agent 凝固剂
settle contour 沉降等值线
settlement 沉降，地陷
settlement measurement 沉降测量
settlement plate 沉降板
settlement stress 沉降应力
settlement velocity 沉降速度
settlement action 沉降作用
settlement calculation by specification
　　规范沉降计算法
settlement calculation depth 沉降计算深度
settlement correction factor 沉降校正系数
settlement curve 沉降曲线
settlement forecast 沉降预测
settlement gauge 沉降测量仪，沉降仪
settlement joint 沉降缝
settlement observation 沉降观测
settlement of pile group 群桩沉降量
settlement of single pile 单桩沉降量
settlement of subsoil 地基沉降量
settlement of compacted soil 土体压后沉降量
settlement of embankment 路堤沉陷
settlement of fill 填土沉陷
settlement ratio 沉降比

settlement ratio of stability 群桩沉降比
settling chamber 沉降室
settling curve 沉降曲线
settling of support 支座下沉
settling rate(velocity) 沉降速度
severe damage 严重破坏
shaft 竖井身
shaft mouth 竖井井口
shaft resistance 桩侧阻
shaft station 井底车场
shaft well 竖井
shaft-sinking [caisson works]
　　　　　打/挖竖井[沉箱工程]
shaking table test 震动台试验
shale 页岩
shale ceramicite 页岩陶粒
shallow arch 扁拱
shallow compaction 浅层压实
shallow cut/dredging 欠挖
shallow failure 浅层滑坡
shallow fill 低填土
shallow footing 浅埋基础
shallow foundation 浅基础
shallow freezing 表层冻土
shallow ground stabilization 表层地基加固法
shallow ground water 浅层地下水
shallow grouting 浅层灌浆
shallow hole blasting 浅孔爆破
shallow hole hydraulic blasting 浅孔液压爆破
shallow percolation 浅层渗透
shallow shaft 浅井
shallow sinking piles with bores 浅层钻孔沉桩
shallow treatment 浅层处理
shallow tunnel 浅埋隧道

shallow well injection 浅井注水
shallow-buried tunnel 浅埋隧道
shallow-focus earthquake 浅源地震
shaly clay 页岩质黏土
shape factor of foundation 基础形状系数
shape factor of section 截面形状系数
shape function 形函数
shape influence coefficient of section shear deformation 截面剪切变形的形状影响系数
shape of hydrograph 水文过程线形式
shape of molecular chain 分子链形态
sharing part of the load 载荷分量
sharp aggregate 多角骨料
sharpness 曲度
shatter rock 震裂岩石
shattered zone 碎裂带
shattering effect 破坏效应
shear 剪切
shear apparatus 剪切仪
shear area 剪切面积
shear band 剪切带
shear box test 剪切盒试验
shear capacity 剪切承载力
shear capacity of inclined section 斜截面承载力
shear carrying capacity 承剪能力
shear center 剪力中心
shear compression failure 剪压破坏
shear compression ratio 剪压比
shear contraction and dilation 剪缩剪胀性
shear crack 剪切裂缝
shear cracking load 剪力开裂荷载
shear deformation 剪切变形
shear deformation of structure plane 结构面的剪切变形

shear diaphragm action 受剪面层作用
shear distortion 剪塌
shear elasticity 剪切弹性模量
shear failure 剪切破坏
shear failure model 剪切破坏模式
shear fissure 剪切裂纹
shear flow 剪力流
shear force diagram 剪力图
shear fracture 剪裂
shear friction 剪摩擦
shear joint 剪切节理
shear lag 剪力滞后
shear lag effect 剪力滞后效应
shear load 剪切荷载
shear locking 剪切闭锁
shear modulus 剪切模量
shear plane 剪切面
shear plate 剪力板
shear rupture of rock 岩石的剪切破坏
shear slant to grain 斜纹受剪
shear slide 剪切滑塌
shear slip 剪切滑动
shear stiffness 抗剪刚度，抗剪劲度
shear strain 剪应变
shear strain rate 剪切应变速率
shear strength 抗剪强度
shear strength of frozen soil 冻土抗剪强度
shear strength of inclined section 斜截面抗剪强度
shear strength of interlayered weak surface
   层间软弱面强度
shear strength of repeated swelling shrinkage
   反复胀缩强度
shear strength of rock 岩石抗剪强度
shear strength of rock mass 岩体抗剪强度

shear strength of soil 土的抗剪强度
shear strength of structure plane 结构面抗剪强度
shear strength parameter 抗剪强度参数
shear strength parameter of soil
    土的抗剪强度参数
shear stress 剪应力
shear stress difference method 剪应力差法
shear test 剪切试验
shear test for rock 岩石剪切试验
shear theory 剪切理论
shear type drill boom 剪式钻臂
shear viscosity 剪切黏度
shear wall 剪力墙
shear wave velocity of soil 土的剪切波速
shear zone 剪切区
shear-bearing capacity 抗剪能力
shear-bending crack 弯剪裂缝
sheared crack 剪切裂缝
shear-friction theory 剪切摩阻理论
shearing 剪切
shearing failure 剪切破坏
shearing force 剪切力
shearing force diagram 剪切力图
shearing limit 剪切极限
shearing modulus of elasticity 抗剪弹性模量
shearing moment 剪切力矩
shearing resistance 抗剪力
shearing section 剪切断面
shearing strain 剪应变
shearing strength 剪切强度
shearing stress 剪应力
shearing stress trajectory 剪应力轨迹
shearing test 剪切试验
shearing type deformation 剪切型变形

shearing vibration 剪切震动
shear-range 剪切幅度
shear-transfer device 剪力传递装置
sheet jointing 片状节理
sheet-pile 板桩
sheet pile anchorage 板桩锚固
sheet pile cut-off wall 板桩防渗墙，板桩截水墙
sheet pile structure 板桩结构
sheet pile wall 板桩墙
sheet pile-braced cuts 板桩围护
sheet piling support 板桩支撑
sheet retaining structure 板式挡土结构
sheeting 板桩墙
sheeting driver 打桩机
sheeting planks of dam 地面护坡，护堤板桩
sheeting/sheet-pile cofferdam 板桩围堰
sheet-pile bulkhead 板桩岸墙
sheet-pile bulkhead with anchor-wall support
 有锚定墙支撑的板桩堤岸
sheet-pile bulkhead with batter pile support
 有斜桩支撑的板桩堤岸
sheet-pile curtain 板桩帷幕
sheet-pile cut-off wall 板桩截水墙
sheet-pile enclosure 板桩围堰
sheet-pile levee 板桩堤
sheet-pile retaining wall 板桩式挡土墙
sheet-pile wall 板桩墙
Shelby tube sampler 谢尔贝薄壁取样器
shell foundation 壳体基础
shelled concrete pile 有套管的混凝土挡土墙
shelter 工棚，风雨棚，遮掩物
Shen Zhujiang three yield surface method
 沈珠江三重屈服面模型
shield 盾构

shield construction 盾构施工法
shield driven 盾构掘进
shield driving method 盾构法
shield engineering 蒙蔽工程
shield excavation 盾构掘进
shield faced with grid 网格式盾构
shield jack 盾构千斤顶
shield method 盾构法
shield sensitivity 盾构灵敏度
shield shell 盾壳
shield steering adjustment 盾构纠偏
shield tail 盾尾
shield tunnel 盾构法隧道
shield tunneling machine 盾构法挖掘机
shield tunnelling method 隧道盾构施工法
shield with balanced earth pressure
　　　　土压平衡式盾构
shield with compressed air 全压式盾构
shield with partial air pressure 局部气压盾构
shield-driven tunnel 盾构式掘进隧道
shield-driven tunneling 用盾构法掘进隧道
shielding with air pressure 气压盾构法
shingle 卵石
shiplap sheet piling 搭叠板桩
shock wave 冲击波
shooting 爆破
shore cutting 岸边冲刷
shore deposit 岸边淤积
shoring of foundation 基础支撑
shoring of trench 槽孔支撑
shoring work 支撑工程
short cylindrical shell 短圆柱壳
short post 短支柱
short term loading test 短期荷载试验

short-term transient load 短期瞬时荷载
shot 爆破,放炮
shot hole 炮眼
shot rock 爆破的岩石
shotcrete 喷射混凝土
shotcrete admixture 喷射混凝土外加剂
shotcrete and rock bolt 喷锚法
shotcrete machine 喷浆机
shotcrete support 喷射混凝土支护
shot-drilled shaft 冲击钻打的竖井
shoulder curb 路边缘石
shoulder line 路肩线
shoulder pitch 路肩斜度
shove joint in brickwork 挤浆砌砖法
shovel car 铲车
shrink mark 缩沉
shrinkable aggregate 收缩集料
shrinkage allowance 允许收缩量
shrinkage and tension joint 伸缩缝
shrinkage coefficient 收缩系数
shrinkage crack 收缩裂缝,缩缝
shrinkage curvature 收缩曲率
shrinkage curve 收缩曲线
shrinkage index 缩性指数
shrinkage limit 缩限
shrinkage limit test 缩限试验
shrinkage modulus 收缩模量
shrinkage ratio 收缩比
shrinkage settlement 收缩沉降
shrinkage strain 收缩应变
shrinkage stress 收缩应力
shrinkage test 收缩试验
shrinkage-creep relationship 收缩-徐变关系
shrinkage-induced cracking 收缩开裂

shut-off water 止水
side abutment pressure 边墩压力
side bend specimen 侧弯试件，侧弯试样
side bend test 侧弯试验
side camber 侧边弯曲
side compression 侧压力
side constraints 界限约束
side cutting 侧向开挖
side ditch 侧沟，边沟
side drain 边沟，路边排水沟
side drift 平硐，平巷
side drift method 侧壁导坑法
side elevation 侧立面，侧面图
side excavation 侧向开挖
side force 侧力
side friction 侧面摩擦
side friction resistance of pile 桩侧摩擦力
side load 侧向负荷
side loading 侧面加载
side of foundation pit 基坑侧壁
side piling 边桩
side pressure 侧压力
side skid 侧向滑移
side slope 边坡
side slope grade 边坡坡度
side slope coefficient 边坡系数
side span 边跨，边孔
sidesway moment 侧移弯矩
sidesway stiffness 侧移刚度
sidetrack horizontal well 侧钻水平井
sidewall core 井壁岩芯
sidewall drill 侧壁钻机
sideway skid resistance 侧向抗滑力
sideways force 横向力

sidewise cut 侧面掏槽
sieve analysis 筛析
sieve classification 筛分分级
sieve curve 筛分曲线
sieve fraction 筛分粒度级
sieve mesh 筛孔
sieve method 筛分法
sieve net 筛网
sieve residue 筛余物
sieve shovel 筛铲
sieve size 筛分粒度
sieve sizing 筛分分级
sieve test 筛分试验
sieve-aperture size for sediment 泥沙筛孔粒径
sifted sand 过筛砂
signature turbulence 特征紊流
silent pile driver 无声打桩机
silent piling 静力压桩
silica 硅石
silica cement 硅土水泥砂浆
silica fume concrete 硅粉混凝土
silicarenite 石英砂岩
silicareous ooze 硅质软泥
silicate injection 硅酸盐注浆
silicification 硅化法
silicification grouting 硅化灌浆
sill 基石，粉土
silt 粉土
siltstone 泥岩，粉砂岩
silty clay 粉质黏土，亚黏土
silty sand 粉砂
simple cement 无掺料水泥
simple shear 单剪
simple shear apparatus 单剪仪

simple shear test 单剪试验
simplex method 单纯形法
Simpson formula 辛普公式
simulated test 模拟试验
simultaneous blasting 齐发爆破
simultaneously grouting 盾构同步压浆
single anchored sheet pile 单锚式板桩
single hole method 单孔法
single hole open caisson 单孔沉井
single line grout curtain 单排灌浆帷幕
single non-central load 偏心集中荷载
single pile 单桩
single row hole open caisson 单排孔沉井
single sized aggregate 均一粒径骨料
single-grained structure 单粒结构
single-shear test 单面剪切试验
single-shot grouting 单液灌浆
sinking by water filling 灌水下沉
sinking of bore hole 钻孔,凿井
sinking of filling 填土沉陷
sinking of open caisson 沉井下沉
sinking of open caisson on filled up island
　　　筑岛沉井
sinking pile by water jet 水力沉桩
sinking speed 沉降速度
sinking vertical shaft by convention 立井普通掘进
siphon pipe 虹吸管
site exploration 场地勘查
site grading 场地平整
site intensity 场地烈度
site investigation 工程地址勘查
site laboratory 工地试验室
site load 施工荷载
site slope of excavation 挖方边坡

site strength 实测现场强度
site survey 现场踏勘，坝址勘测
site trial 实地试验，工地试验
size distribution curve 粒度分布曲线
size division 粒度级别
size effect 尺寸效应
size factor 粒度系数
size grading 粒径级配
size of load 载荷
size of mesh 筛号
size of screen mesh 筛孔尺寸
skeleton curve 骨架曲线
Skempton's ultimate bearing capacity formula 斯肯普顿极限承载力公式
sketch 草图
sketch drawing 绘制草图
sketch for the colluvial soils 崩积土图样
skid force 滑移力
skid pile-driver 滑动打桩机
skid-resistance 抗滑性
skid-resistance value 抗滑系数
skid-resisting capacity 抗滑能力
skin friction 表面摩擦，表面摩擦力
skin friction of pile 桩侧摩阻力
skin friction pile 摩擦桩
skin stress 表层应力
skin friction of driven pile 打入桩摩阻力
skin-friction force 表面摩擦力
skip grading curve 不连续级配曲线
skip winder 箕斗提升机
slab culvert 盖板式涵洞
slab foundation 板式基础
slack variables 松弛变量
slacktip 崩塌

slag 矿渣
slag aggregate 矿渣集料
slag cement 矿渣水泥
slag expansive cement 矿渣膨胀水泥
slag fill 矿渣填料
slag gypsum cement 石膏矿渣水泥
slag Portland cement 矿渣波特兰水泥
slag-sulfate cement 硫酸盐矿渣水泥
slake-durability index of rock 岩石的耐崩解性指标
slaking 湿化
slaking characteristic of rock 岩石的崩解性
slaking of soil 土的崩解性
slaking property 水解性
slaking test 湿化试验
slate 板岩
sleeve pipe grouting 套管法灌浆
slender column 细长柱
slender ratio 长细比
slice method 条分法
slickenside 擦痕面
slide deformation 滑坡变形
slide-resistant pile 抗滑桩
sliding stability 滑动稳定性
slip coefficient 抗滑移系数
slip line 滑动线
slip surface 滑动面
slip zone 滑动带
slope 土坡
slope erosion 边坡冲刷
slope failure 斜坡崩塌，山泥倾泻，滑坡
slope flaking 边坡剥落
slope inversion analysis 边坡反分析
slope level 测斜仪

slope monitoring 边坡监测
slope protection 护坡
slope protection works 护坡工程
slope stability 斜坡稳定性
slope stability analysis 斜坡稳定性分析
slope stabilization 边坡加固
slope stake 边坡桩
slope surface erosion 坡面冲蚀
slope toe 坡脚
slope toe wall 坡脚墙
slope treatment works 斜坡整治工程
slope wash 坡积土,坡积物,斜坡侵蚀
sloping foundation 倾斜基础
sloping core 斜墙
slough 坍塌
sloughing shale 坍塌页岩
slow shear test 慢剪试验
sludge pump 排泥泵
slump 坍落度
slump belt 崩落地带
slump block 坍塌块体
slump cone 坍落度筒
slump consistency test 坍落稠度试验
slump constant 坍落度
slump loss 坍落度损失
slump test 坍落度法
slumping 崩滑
slurry 泥浆
slurry bored pile 泥浆护壁钻孔灌注桩
slurry coat method 护壁泥浆法
slurry cut-off 泥浆防渗墙
slurry equipment 拌浆设备
slurry explosives 浆状炸药
slurry for preventing collapse of borehole

泥浆护壁
slurry of ceramic 陶瓷泥浆
slurry pump 泥浆泵
slurry purification system 泥浆净化设备
slurry sampler 泥浆取样筒
slurry stable wall 泥浆固壁
slurry transport 浆料输送
slurry trench method 泥浆槽法(地基防渗处理)
slurry trench wall 泥浆防渗墙,地下连续墙
small strain 小应变
small-size mechanical rammer 小型夯土机
smectite 蒙脱石
smooth surface blasting 光面爆破
soakage pit 渗水井
soaking line 浸润线
soaking surface 浸润面
socket foundation 杯口基础
sod revetment 草皮护坡
sodium bentonite mud 钠膨润土泥浆
sodium silicate 水玻璃
soft clay 软黏土
soft clay ground 软土地基
soft flow 高流动性
soft formation 软岩层
soft foundation 软土地基
soft ground floor 柔性底层
soft rock 软岩
soft rock mass 松软岩体
soft rock strata 软弱夹层
soft soil 软土
soft soil foundation 软土地基
softening 软化
softening agent 软化剂
softening coefficient of rock 岩石软化系数

softening index 软化系数
softening property of rock 岩石软化性
soil 土
soil absorption capacity 泥土渗水能力
soil air 土壤空气
soil anchor 土层锚杆
soil and rock testing 土及岩石试验
soil and water conservation 水土保持
soil arch effect 土拱效应
soil auger 土样钻取器
soil bank 土堤
soil bearing capacity 土的承载能力
soil bearing value 土承载力值
soil binder 土的胶结物
soil cement 掺土水泥
soil cement processing 水泥土加固法
soil classification 土壤分类
soil cofferdam 土围堰
soil colloid 土胶体
soil column 土桩
soil column stabilization method 土桩加固法
soil compactibility 土壤密实度
soil compaction 土的压实
soil compaction by lime-soil pile
    灰土挤密桩加固法
soil compaction by stone pile 挤密碎石桩加固法
soil compaction pile 土桩
soil compaction pile method 挤密土桩法
soil composition 土的组成
soil concrete 黏土混凝土
soil conservation 水土保持
soil consolidation 土固结
soil cutting 切土
soil damping 土的阻尼

soil dynamics 土动力学
soil evaporation 土壤蒸发
soil exploration 土质勘探，土质调查
soil fabric 土的组构，土的结构
soil fines 土的细粒部分
soil flow 流土
soil freezing prevention 土壤防冻
soil grain 土粒
soil grain distribution test 土壤粒径分布试验
soil group 土类
soil improvement 土质改良，地基加固
soil in place(situ) 现场土
soil lab 土工试验室
soil layer response calculation 土层反应计算
soil legend 土的图例
soil lines of equal pore pressure 土内等孔隙水压线
soil liquefaction 土的液化
soil map 土分布图
soil mass 土体
soil mechanics 土力学
soil mineralogy 土矿物学
soil moisture stress 土中水分应力
soil moisture 土中水
soil nail 泥钉，土钉
soil nailing 土钉
soil nailing wall 土钉墙
soil name 土名
soil packer 压土器
soil parameter 土工参数
soil particle 土粒
soil pattern 土样
soil permeability 土的渗透性
soil pile 土桩

soil piping 土的管涌
soil porosity 土孔隙率［多孔性］
soil pressure 土压力
soil profile 土体剖面
soil properties 土的性质
soil reinforcement 土的加筋法
soil sack cofferdam 土袋围堰
soil sample 土样
soil sample analysis sieve 土样分析筛
soil sample barrel 取土筒
soil sample box 取样盒
soil sampler 取土器
soil sampling 取样
soil sampling tube 取土管
soil section 土层剖面
soil separator 土选分机
soil series 土系
soil shifter 推土机
soil skeleton 土骨架
soil slope 土坡
soil sounding device 土触探装置
soil stabilization 土体加固法
soil stabilization method by sheet pile 板桩加固法
soil strata 土层
soil stratification 土的层理，土的分层
soil strengthening 土的加固
soil structure 土的结构
soil suction 土的吸力，土的负孔隙压力
soil suction potential 土吸力势
soil survey 土质调查
soil test 土工试验
soil texture 土的构造
soil water 土中水
soil water movement 土中水分运动

soil water retension characteritics 土壤水分特征
soil water retension curve 土壤持水曲线
soil water tension 土的水压力
soil wedge 土楔
soil/rock support 岩土支撑面
soil-cement material 水泥土
soil-cement processing 水泥土加固法
soil-sampling device 打入式开缝管取土器
soil-structure interaction 土-结构相互作用
soldier beam 立桩
soldier pile 排桩，竖桩，立桩
solid map 岩石分布图
solid mechanism 固体力学
solid raft 实体筏基
solid rock 坚石，坚岩
solidification grouting 硅化灌浆
solidity ratio 硬度比
solubility 溶解度
solubility curve 溶解度曲线
soluble clay 可溶性黏土
soluble salt test 可溶性盐试验
soluble segregation 溶质偏析
solution cave 溶蚀洞穴
solution cavity 溶洞
solution channel 溶蚀槽
solution channel spring 溶洞泉
solution concentration 溶液浓度
solution groove 溶沟
solution grout 溶液灌浆
solution injection 溶液灌浆
solution suction 溶质吸力
sonic coring 超声波验桩法
sonic pile driver 声波打桩机，振动打桩机
sonic prospecting 声波法探测

sound cement 安全水泥
sound unaltered rock 坚固完整岩石，坚固新鲜岩石
sounding method 触探法
sounding well 探井
spacing between blast holes 炮眼间距
spall 破碎
spallability 可破碎性
spalling 劈裂
special foundation 特殊基础
special method of shaft sinking 立井特殊掘进
special soil 特殊土
special-shaped cast-in-place pile 异形灌注桩
specific absorption 比吸水量
specific body force 单位体力
specific charge 单位体积装药量
specific creep 比徐变
specific density of solid particles 土粒相对密度
specific density(gravity) 相对密度(比重)
specific energy 比能
specific gravity of soil particle
　　　　　土粒相对密度(比重)
specific gravity capacity 比阻尼容量
specific gravity test 比重(相对密度)试验
specific grout absorption 单位吸浆量，比吸浆量
specific penetration resistance 比贯入阻力
specific strength 比强度
specific surface 比表面积
specific surface area of particle 颗粒比表面积
specific surface coefficient of structure
　　　　　结构表面系数
specific surface energy 比表面能
specific surface tension 比表面张力
specific surface test 比表面积试验

specific value of stress 应力比
specific viscosity 比黏度
specific volume 比容
specific water absorption 单位吸水量
specific weight for fluid 流体重度
specific yield 给水度
specified penetration 指定贯入度
specimen 试件
speed cement 快速水泥
speed of pressuring pile 压桩速度
speedy drivage 快速掘进
Spencer method 斯宾赛法
spent shot 无效爆破
spherical particles 球状颗粒
spherical stress 球应力
spile 插板
spillway tunnel 泄洪隧洞
spiral cut 螺旋掏槽
splinter 裂片
split 劈裂
split tensile strength 劈裂抗拉强度
split tensile test 劈裂抗拉试验
split(splitting) test 劈裂试验,巴西试验
splitting strength 劈裂强度
splitting tensile strength 劈裂抗拉强度
spoil 弃方(土),矸石
spongy rock 多孔岩石
spoon penetration test 土勺灌入试验
spout of stock 喷浆
sprayed concrete 喷射混凝土
sprayer 喷头
spread angle of pressure in subsoil
　　　地基压力扩散角
spread footing 扩展基础

spring cone penetrometer 弹簧锥贯入度计
square footing 方形基础
square spread footing 方形独立基础
squeeze test 挤压实验
stability 稳定性
stability against overturning 抗倾稳定
stability against sliding 抗滑稳定
stability analysis 稳定分析
stability analysis of diaphragm trenches 槽壁稳定性验算
stability analysis of rock slope 岩坡稳定分析
stability analysis of slope 土坡稳定分析
stability assessment 稳定性评估
stability augmentation system 稳定性加强系统
stability curve 稳定性曲线
stability equation 稳定方程
stability factor 稳定(性)系数
stability factor against overturning 倾覆稳定系数
stability investigation 稳定性勘察
stability number 稳定数
stability of float caissons 浮式沉井稳定性
stability of foundation rock 岩基稳定性
stability of foundation soil 基础稳定性
stability of reservoir slope 水库岸坡稳定性
stability of retaining wall 挡土墙稳定性
stability of rock mass 岩体稳定性
stability of subsoil 地基稳定性
stability of surrounding rock 围岩稳定性
stabilization by densification 挤密加固
stabilization by vibroflotation 振冲加固
stabilization method by mixture 拌合加固法
stabilization of landside 滑坡加固
stabilization of superficial subgrade 浅层地基加固法

stabilization of deep subgrade	深层地基加固
stabilizing fluid	固壁泥浆
stabilizing grout	固结灌浆
stabilizing grout	稳定浆液
stabilizing moment	稳定力矩
stable borehole	井壁稳定
stable crack growth	稳定裂纹扩展
stable rock bed	稳定岩基
stack filling	堆填法
stage cementing technology	分段固井技术
stage grouting	分期灌浆
staggered arrangement of piles	桩的梅花式排列
staggered piling	打梅花桩，错列打桩
stake out	打桩
stanchion base	支柱基础
stand pipe	孔口管
standard frost depth	标准冻结深度
standard penetration test(SPT)	标准贯入试验
standard sand	标准砂
standard screen(sieve)	标准筛
standard specimen for rocks	岩石标准试件
standpipe piezometer	竖管式测压计
state of elastic equilibrium	弹性平衡状态
state of limit equilibrium	极限平衡状态
state of loading	载荷状态
state of plastic equilibrium	塑性平衡状态
state of saturation	饱和状态
state of stress	应力状态
state parameters	状态参数
state variable	状态变量
static analysis(shallow foundation)	静定分析法(浅基础)
static cone penetration test	静力触探试验，静锥触探试验

static cone sounding 静力触探
static formula of pile 桩的静力公式
static friction 静力摩擦
static head 静水压
static load 静荷载
static load test 静载荷试验
static load test of pile 单桩竖向静载荷试验
static penetration test 静力初探试验
static pile loading 静力压桩
static pile pressure-extract machine 静力压拔桩机
static point resistance 静力触探探头阻力
static pressure 静压力
static sounding 静力触探
static test 静力试验
static test load 静态试验载荷
static water level 静水位
static water pressure 静水压力
stationary flow 稳态流
stationary flow field 稳态流场
stationary mortar mixer 固定式灰浆搅拌机
stationary stream 稳态流，稳定流
stationary-piston sampler 固定活塞式取样器
statistic soil mechanics 统计土力学
statistical corrections for bearing capacity value
　　承载力值的统计修正
statistical damage theory 统计损伤理论
statistical hydrology 统计水文学
stay pile 拉索桩
steady creep 稳定蠕变
steady flow 恒定流
steady secondary ground pressure 二次地压
steady seepage(flow) 稳定渗流
steady water table 稳定水位
steady-state creep of rock 岩石稳定蠕变

steel casing 铸钢井壁
steel open caisson 钢沉井
steel open caisson with double-shell 双壁钢沉井
steel pile 钢桩
steel pipe pile 钢管桩
steel pneumatic caisson 钢沉箱
steel reinforcement cage 钢筋笼
steel sheet pile 钢板桩
steel sheet pile cofferdam 钢板桩围堰
steel tube pile 钢管桩
steep dip 陡倾
steep slope 陡坡
stepped footing 挡土墙台阶基础
stepped foundation 阶梯形地基
stereographic projection 赤平投影
S-test 慢剪试验
stick force 外附力
sticky limit 粘限
stiff clay 硬黏土
stiffening 固化，硬化
stiffness 刚度，劲度
stiffness factor 刚度系数
stiffness ratio 刚度比
stirring machine 搅拌机
stone 碎石土
stone column 碎石桩
stone drift 岩石平巷
stone footing 毛石基础
stone material 石料
stone revetment 干砌石护坡
stone-concrete footing 毛石混凝土基础
stony soil 石质土
storage coefficient 储水系数

straight cement mortar 纯水泥浆
strain aging 应变时效
strain and stress in shell 壳体的应变和应力
strain cell 应变计
strain control 应变控制
strain control triaxial compression apparatus 应变控制式三轴压缩仪
strain controlled test 应变控制试验
strain coordination factor 应变协调因子
strain corresponding to the maximum stress 峰值应力应变
strain crack 应变裂缝
strain cracking 应变开裂
strain curve 应变曲线
strain deviator tensor 应变偏斜张量
strain diagram 应变图
strain energy 应变能
strain forces 变形力
strain gauge 应变仪
strain harding 应变硬化
strain harding law 加工硬化定律
strain lag of added concrete 应变滞后
strain meter 应变仪
strain rate(ratio) 应变率
strain rosettle 应变花式应变仪
strain softening 应变软化
strain space 应变空间
strain spherical tensor 应变球形张量
strain state at a point 一点应变状态
strain surface 应变曲面
strain tensor 应变张量
strain vector 应变向量
strain-controlled testing method 可控应变试验法

strainer well 过滤井
strainer-gage type displacement transducer
 应变式位移传感器
strain-hardening behaviour of rock
 岩石的应变硬化性
strain-hardening stage 应变强化阶段
strain-softening behaviour of rock
 岩石的应变软化性
strand 预应力钢缆,钢绞线
strap footing 条形基础
strata control 地层控制
stratification 层理,分层现象,成层作用
stratification foundation 层状地基
stratification plane 层理面
stratified random sampling 分层取样
stratified relief 层状地貌
stratified rock 成层岩
stratified soil 层状土,成层土
stratified structure 层状结构
stratigraphic 地层的
stratigraphic cross section 地层横剖面
stratum 地层
stratum water 层状水
straw and earth cofferdam 草土围堰
stream abstraction 河流合流作用
stream action 河流作用
stream bank stabilization
 河岸稳定,河岸加固
stream closure 截流
stream closure hydraulics 截流水力学
stream energy 河流能量
stream erosion 河流冲刷
stream fluting 河流槽蚀
stream function 流函数

stream load 河流泥沙量
strength 强度
strength analysis 强度分析
strength anisotropy index
    强度各向异性指标
strength calculation 强度计算
strength ceiling 极限强度
strength characteristic 强度特征值
strength class 强度分级
strength condition 强度条件
strength contour graph 等强度线图
strength criterion of slope safety
    边坡安全性标准
strength curve surface of rock 岩石强度曲面
strength envelope 强度包线
strength factor of material 材料强度系数
strength grade 强度等级
strength grading of mortar 砂浆强度等级
strength loss ratio of rock 岩石强度损失率
strength model 强度模型
strength of axially-loaded member
    轴心受力构件强度
strength of compression 抗压强度
strength of rock 岩石强度
strength of rock mass 岩体强度
strength reduction factor 强度降低系数
strength safety coefficient 强度储备系数
strength test 强度试验
strength theory 强度理论
strength theory of rock 岩石强度理论
strength to density ratio 强度-密度比
strength under sustained load
    长期荷载作用下的强度
strength-aging relationship 强度-龄期关系

strength-deformation characteristic
　　　强度变形特性
strength-density ratio　强度-密度比
strengthen layer　补强层
strengthened stage　强化阶段
strengthening　补强，加固
strengthening by struts　支顶加固
strengthening grouting　补强灌浆
strengthening measure for earthquake resistance
　　　抗震加固措施
strengthening method with prestressed brace bar
　预应力撑杆加固法
strengthening of structure under loading
　　　带负荷加固法
stress　应力，压力
stress accommodation　张紧夹具
stress along bolt　锚杆应力
stress analysis　应力分析
stress and strain response　应力及应变反应
stress beyond yield strength　超过屈服强度的应力
stress circle　应力圆
stress concentration　应力集中
stress concentration diffusion　应力集中的扩散
stress concentration factor　应力集中系数
stress constraint　应力约束
stress controlled test　应力控制试验
stress control triaxial compression apparatus
　　　应力控制式三轴压缩仪
stress corrosion　应力腐蚀
stress corrosion cracking　应力腐蚀开裂
stress crack　应力裂缝
stress deviatoric tensor　应力偏斜张量
stress diagram　应力分布图
stress dispersion　应力扩散

stress distribution 应力分布
stress due to hammer driving 锤击应力
stress due to temperature difference
　　　温差应力
stress excess of tensile steel 应力超前
stress field 应力场
stress for temporary loading 短期荷载应力
stress function 应力函数
stress function of bending 弯矩应力函数
stress function of torsion 扭转应力函数
stress gauge 应力计
stress gradient 应力梯度
stress history 应力历史
stress in original rock 原岩应力
stress in prestressing tendon at transfer or anchorage 传力锚固应力
stress in soil mass 地基应力，土体中的应力
stress in the surrounding rock 围岩应力
stress induced fracture 应力引起的断裂
stress intensity factor
　　　应力强度因子，应力强度系数
stress intensity factor calculation method
　　　应力强度因子计算方法
stress level 应力水平
stress method 应力解法
stress path 应力路径
stress path method 应力路径法
stress range 应力范围
stress ratio 应力比
stress ratio of liquefaction 液化应力比
stress recovery method 应力恢复法
stress redistribution 应力重分布
stress regulation 应力调整
stress relax meter 应力松弛仪

stress relaxation 应力松弛
stress release 应力解除，应力释放
stress releasing borehole 应力解除钻孔
stress relief method 应力解除法
stress relieving 应力解除
stress repeation 应力反复
stress space 应力空间
stress spectrum 应力谱
stress spherical tensor 应力球形张量
stress state at a point 一点应力状态
stress strength factor 应力强度因子
stress surface 应力曲面
stress tensor 应力张量
stress to strain ratio 应力应变比
stress vector 应力向量
stress wave in soils 土中应力波
stress-controlled testing method
　　　　可控应力试验法
stress-deformation characteristic
　　　　应力变形特征
stressed shell construction 张壳构造
stressed skin action 应力蒙皮作用
stress-equilibrium method 应力平衡法
stresses due to hammer driver 锤击应力
stress-freezing effect 应力冻结效应
stressing method 先张拉法
stressing process 张拉过程
stress-producing force 引起应力的外力
stress-ratio coefficient of original rock
　　　　原岩应力比值系数
stress-ratio method 应力比法
stress-release channel 应力解除槽
stress-strain curve 应力-应变曲线
stress-strain curve in tension

受拉应力-应变曲线
stress-strain diagram 应力-应变图
stress-strain relationship of mortar
　　砂浆的应力-应变关系
stress-strain relationship of soil under cyclic loading
　　循环荷载下土的应力-应变关系
stretch elongation 拉伸变形
stretch forming 张拉成形法
stretcher strain 拉伸变形
stretching 延伸，张拉
stretching force 张力
stretching in batches 成组张拉
stretching process 张拉过程
stretching strain 张拉应变
stretching stress 张拉应力
stretching wire 张拉钢筋
striated pebble 擦痕卵石
striated rock surface 擦痕面
striation 擦痕
striction strain 颈缩应变
strike direction 走向
strike-slip fault 走滑断层，平移断层
strip footing(foundation) 条形基础
strip foundation under walls 墙下条形基础
strip load 条形荷载
structural block 结构体
structural characteristic 结构特征
structural landform 构造地貌
structural plane 结构面
structural strength of soil 土的结构强度
structural types of rock mass 岩体结构类型
structure of rock 岩石构造
structure-foundation-soil interaction analysis
　　上部结构—基础—地基共同作用分析

subaerial deposit 陆上沉积物
subangular particles 次棱角土粒
subbase 地基
subgrade 路基
subgrade coefficient method 基床系数法
submarine soil 海底土
submerged density of soil 土的浮密度
submerged unit weight of soil
      土的浮表观密度（浮容重）
submerged weight density of soil 土的浮重度
subsidence 沉淀，下沉
subsidence trough 塌陷坑
subsidiary shaft 副井
subsidiary shot hole 辅助炮眼
subsiding by matching weights 配重下沉法
subsoil 天然地基
subsoil permissible load 天然地基允许荷载
subsoil water 潜水
substitution 置换
substratum 下卧层
subsurface exploration 地基勘探
subsurface runoff 地下径流
subsurface seepage flow 地下渗流
subsurface water 潜水
sullage 沉积的淤泥
sump 贮(油、液)槽
sunk caisson(well)foundation 沉井基础
sunken tube method 沉管法
superficial compaction 表面压实
superimposed soil 上覆土
superimposed stress 附加应力
superload 超载，附加荷载
support ped 支撑垫板
supporting course 持力层

surcharge 超载
surcharge preloading 超载预压
surface compaction 表层压密法
surface elastic wave 弹性表面波
surface flow 表流，地表径流
surface force 表面力
surface friction 表面摩擦，表面摩擦力
surface grouting 接触灌浆
surface settlement plate 表面沉降板
surface soil stabilization 浅层土加固
surface subsidence 地面塌陷
surface tension 表面张力
surface texture 表面纹理，表面构造
surface treatment 表面处理
surface water 地表水
surface wave velocity method 表面波法
surface wave test 表面波试验
surficial soil 表土
surrounding rock 围岩
surrounding rock pressure 围岩压力
surrounding rock stress 围岩应力，二次应力
swamp soil 沼泽土
S-wave 剪切波，S 波
Swedish circle method 瑞典圆弧法
swelling force 膨胀力
swelling index 回弹指数
swelling of floor 地鼓
swelling rate test 膨胀率试验
swelling ratio 膨胀率
swelling soil 膨胀土
swelling strain index 膨胀应变指数
swelling test 膨胀试验
syenite 正长岩
sympathetic detonation 感应起爆

synchronous pile 成孔灌注同步桩
syncline 向斜

T

taconite 铁遂石，铁英岩
tail 盾尾
tailings 尾矿砂
tailings dam 尾矿坝
talc 滑石，云母
talik 层间不冻层
talus 坝脚抛石
talus fan 冲积扇，岩屑扇
talus wall 单侧坡面墙
tamp 填塞
tamped backfill 夯实回填土
tamper 夯土机
tamping 夯实
tamping area(range) 夯击面积，夯击范围
tamping energy 夯击能
tamping in layers 分层夯实
tamping speed 打夯速度
tangent bend 双弯曲
tangent curve 正切曲线
tangent friction force 切线摩擦力
tangent modulus 切线模量
tangent modulus of rock 岩石切线模量
tangent modulus of soil 土的切线模量
tangent modulus theory 切线模量理论
tangent pile 支护桩
tangent stiffness 切线刚度
tangential force 切向力
tangential frost-heaving stability of foundation

基础切向冻胀稳定性
tangential strain 切应变
tangential stress 切应力
tapered pile 锥形桩
Taylor method 泰勒法
tectonic earthquake 构造地震
tectonic feature 构造特征
tectonic fissure 构造裂缝
tectonic ground crack 构造性地裂缝
tectonic joint 构造节理
tectonic movement 构造运动
tectonic stress 构造应力
tectonic structural surface(structure plane)
　　　　构造结构面
telescope grouting method 套管式灌浆法
telescopic bucket reamer 伸缩式扩孔机
Telford base 大块石基层
Telford pavement 大石块基层碎石路面
Telford road 大石块基层道路
tell-tale 应变杆
temperature action 温度作用
temporary rock support 临时岩体支护
temporary strengthening 临时加固
temporary support 临时支撑
temporary trench support
　　　　临时槽支撑，临时坑壁支撑
temporary water-bearing layer 暂时含水层
tenacious clay 强黏性土
tenacity 黏滞性
tensile adhesion property 拉伸粘结性能
tensile area 受拉区
tensile capacity 受拉承载力
tensile crack 受拉裂缝
tensile curve 拉伸曲线

tensile failure 受拉破坏
tensile force 拉力，张力
tensile fracture 张断面
tensile impact 张力冲击
tensile load 受拉荷载
tensile modulus 拉伸模量
tensile region 受拉区
tensile splitting strength 劈裂受拉强度
tensile strain 拉伸(张)应变
tensile strength 抗拉强度
tensile stress 拉应力
tensile test 拉力测试
tensile yield point 拉力屈服点
tensiometer 张力计
tension 张力，拉力
tension brace 拉撑
tension crack 张裂缝，拉裂缝
tension crack depth on the top of slope
　　　　坡顶裂缝开展深度
tension curve 张力曲线
tension failure 拉断破坏，受拉破坏
tension fissure 张裂缝
tension fracture 拉裂
tension grip 张拉夹具
tension pile 拉力桩
tension strength 抗拉强度
tension stress during driving
　　　　打桩时的张拉力
tension structure 张拉结构
tensional rigidity 抗拉刚度
tensioning by section 分段张拉
tensioning device(equipment) 张拉装置
tensioning force 张力
tensioning procedure 张拉程序

tensometer 张拉计
terminal pressure 终点压力
termination of grouting 灌浆终结
terrace 阶地
terrace wall 梯状挡土墙
terrain 地形,地势,地貌
terrain line 地形线
terra-probe 水下振砂器
Tertiary 第三纪
Terzaghi bearing capacity theory
　　　　太沙基承载力理论
Terzaghi theory 太沙基理论
Terzaghi-Rendulic diffusion equation
　　　　太沙基—伦杜列克扩散方程
Terzaghi's ultimate bearing capacity equation
　　　　太沙基地基极限承载力公式
Terzaghi's consolidation theory 太沙基固结理论
test by coring sample 钻芯法实验
test by injecting water into the grout hole
　　　　压水试验
test cube 立方体试块,试块
test embankment 试验填方,试验堤,碾压试验
test fill 碾压试验填土
test for losses of prestress 预应力损失试验
test loading design 试验加载设计
test on completion 竣工检验
test on site 现场试验
test piece 试件
test pile 试桩
test pit 探井
test result 试验成果
test shaft 试井
test specimen 试件
test trench 探槽

testing equipment 试验设备
testing load 试验荷载
texsol(fiber soil) 纤维土
texture 纹理
texture of rock 岩石结构
thaw collapse 融陷
thaw collapsibility 融陷性
thaw compression test of frozen soil
　　　冻土融化压缩试验
thaw consolidation 冻融固结
thaw depth 融化深度
thaw subsidence factor 融沉因数
thawed soil 融土
the method of flat jack 扁式液压顶法
the method of mortar flake 砂浆片剪切法
the method of point load 点荷法
the thickness of the layer 土层的厚度
theory of yield surface of soil 土的屈服面理论
theory of buckling 屈曲理论
theory of consolidation 固结理论
theory of damping 阻尼理论
theory of elastic thin shell 弹性薄壳理论
theory of elastoplastic model of soil
　　　土的弹塑性模型
theory of flow rule of soil 土的流动规则理论
theory of high packing 紧密堆积理论
theory of stability 稳定理论
theory of strain gardening law of soil
　　　土的加工硬化理论
theory of strength 强度理论
theory of thermoviscoelasticity 热黏弹性理论
theory of vibrating densification 振实理论
thermal consolidation 热固结
thermal cracking 温度裂缝

thermal deformation 热变形
thermal differential analysis 差热分析
thermally induced strain 热生应变
thick grout 浓浆，稠浆
thick wall sampler 厚壁取土器
thickness of compressed layer 压缩层厚度
thickness of compressed layer of foundation 地基压缩层厚度
thickness of site soil layer 场地覆盖层厚度
thin cast-in-situ diaphragm wall 现浇薄膜防渗墙
thin wall sampler 薄壁取土器
thixotropic fluids 触变泥浆
thixotropy 触变性
thixotropy of soil mass 土体的触变
thorough consolidation 完全固结
thread anchorage 螺丝端杆锚具
three dimensional stress 三维应力
three phase diagram 三相图
three-dimensional consolidation settlement 三向变形条件下的固结沉降
threshold analysis 临界分析
throwing blasting 抛掷爆破
thrust 推力
thrust at springer 拱脚推力
thrust curve 推力曲线
thrust force 推力
tie rod 拉杆
tieback 锚杆，锚固
tieback retaining wall 锚定板式挡土墙
tieback wall(anchored wall) 锚杆挡墙
tight fissure 致密裂缝
tight loading 密实装药
till 冰碛土
tilt load 倾斜荷载

tilt monitoring device 倾斜监测装置
tilt stability 倾覆稳定性
tilted stratum 倾斜岩层
tilting correction by earth undercutting 掏槽纠偏
tilting correction by ballast loading 堆载加压纠偏
tilting correction by undercutting and dewatering 降水掏土纠偏
tilting failure 倾覆破坏
tilting moment 倾覆力矩
tiltmeter 倾角量测仪
timber 坑木
timber piles 木桩
timber pneumatic caisson 木沉箱
timber sheet pile 木板桩
time analysis 时间分析
time consolidation curve 时间固结曲线
time constant 时间常数
time effect 时间效应
time factor 时间因子,Tv
time-dependent effects of rock 岩石的时间效应
time-yield 蠕变
tiny crack 微裂缝
tip resistance 桩尖阻力
tjaele 冻土
toe circle 坡脚圆
toe cut 底部掏槽
toe excavation 坡脚开挖
toe of slope 坡脚,坡角
toe wall 趾墙
tolerable settlement 容许沉降量
tolerance 容忍,限差
top cut 顶部掏槽
topographic condition 地形状况

topographic latitude 大地纬度
topographic map 地形图
topographic point 地形点
topographic reconnaissance 地形勘探
topographic survey 地形测量
topography 地形，地势，地貌
toppling failure 倾倒破坏
topsoil 表层土
torpedo sand 粗粒砂
torque 扭矩
torque coefficient 扭矩系数
torque reaction 反作用扭矩
torsibility 耐扭力
torsion shear apparatus 扭剪仪
torsion test for rock 岩石扭转试验
torsional stiffness factor 抗扭刚度系数
torsional buckling 扭转失稳
torsional deformation 扭转变形
torsional force 扭力
torsional rigidity 抗扭刚度
torsional shear test 扭剪试验
torsional shearing stress 扭转应力
torsional strength 抗扭强度
torsional tester 扭转试验机
torsional vibration 扭转振动
torsion-bending ratio 扭弯比
torsion-shear process 扭剪法
torsion-shear ratio 扭剪比
total stress analysis of shear strength
    抗剪强度总应力法
total stress method of stability analysis
    土坡稳定分析的总应力法
total collapse 总湿陷量
total deflection 总变形

total deformation 总变形量
total depth 总钻深
total displacement 总位移量
total heave 总隆起量
total mineralization of groundwater
　　　　地下水总矿化度
total moisture content 总含水量
total porosity of rock 岩石的全孔隙度
total settlement 总沉降量
total stress 总应力
total stress analysis 总应力分析
total stress approach of shear strength
　　　　抗剪强度总应力法
total stress failure envelope 总应力破坏包线
total stress path 总应力路径
total stress strength parameter 总应力强度参数
total tolerance 总容许限度
total water cement ratio 总水灰比
total weight 总重量
transducer 传感器
transfer length for pretension tendon
　　　　预应力钢筋传力长度
transfer length of prestress 预应力传递长度
transformation of stress space 应力空间转换
transformed section 换算截面
transformed section method 换算截面法
transient load 瞬时荷载
transmissivity 导水系数
transported soil 运积土
transverse permeability 横向透水性
trap rock 暗色岩
traveled soil 运积土
treatment by sand-bath 喷砂处理
treatment of negative skin friction 桩负摩阻处理

treatment with short pile 短桩处理
tremie method 导管法
trench 探槽,地沟,沟槽
trench method 堑壕法
trench backfill 沟槽回填
trench cut method 挖沟法
trenching 槽探
Tresca yield condition 特雷斯卡屈服条件
trial boring 试钻
trial driving 试验打桩,试桩
trial heading 探洞
trial hole 探坑
trial pile 试桩
trial piling 试验打桩
trial pit 试坑
trial wedge method 滑楔试算法
triangle cut 三角掏槽
triangular method 三角形法
Triassic 三叠纪
triaxial apparatus 三轴仪
triaxial cell 三轴压力室
triaxial compression strain of rock
    岩石三轴压缩应变
triaxial compression strength of rock
    岩石三轴抗压强度
triaxial compression test 三轴压缩试验
triaxial extension test 三轴拉伸试验,三轴伸长试验
triaxial shear test 三轴剪切试验
triaxial shrinkage test 三轴收缩试验
triaxial state of soil 三轴应力状态
triaxial test 三轴试验
tri-phase soil 三相土
triple-pipe chemical churning process

　　　　三重管旋喷法
troublesome soil　难处理土
true angle of internal friction　真内摩擦角
true cohesion　真黏聚力
true specific gravity　真相对密度(比重)
true strain　真应变
true stress　真应力
true triaxial apparatus　真三轴仪
true triaxial test　真三轴试验
true triaxial testing machine for rock
　　　　岩石真三轴试验机
true value　真值
true viscosity　真实黏度
Tsinghua elastoplastic model
　　　　清华弹塑性模型
tsunami　海啸，地震海啸
tube caisson foundation　管柱基础
tube sampler　管式取样器
tube sinking method　沉管法
tube well drainage　管井排水
tube-immersing(sinking)　管段沉入
tubular colonnade foundation　管柱基础
tubular construction　筒形结构
tubular foundation　管状基础
tubular pile　管桩
tufa cement　凝灰岩水泥
tuff　凝灰岩
tundra　冻土
tundra soils in the arctic area　北极冰沼土
tunnel　隧道
tunnel borer(boring machine)　隧道掘进机
tunnel boring machine method　隧道掘进机法
tunnel construction　隧道施工
tunnel diversion　遂洞导流

tunnel drainage 隧道排水
tunnel drainage during construction 隧道施工排水
tunnel drill 隧道凿岩机
tunnel effect 隧道效应
tunnel erosion 隧道洞蚀
tunnel excavation 遂道开挖
tunnel heading 隧道导坑
tunnel invert 隧道底拱
tunnel lining 隧道衬砌
tunnel profile 隧道截面
tunnel ring 隧道圈
tunnel roof 隧道顶板
tunnel shaft 隧道竖井
tunnel shield 隧道盾构
tunnel spoil 隧道弃方
tunnel support 隧道支撑
tunnel survey 隧道测量
tunnel under pressure 有压隧洞
tunneling 隧道工程
tunneling with pilot pipes 管棚法
turbulence 湍流
turbulent flow 紊流
turf 泥炭
turf protection of slope 草皮护坡
twice grouting method 二次注浆法
twin shear stress yield criterion 双剪应力屈服模型
twist auger 螺旋钻
two dimensional consolidation 二维固结
two-fluid process 双液法(化学灌浆)
two-parameter foundation model 双参数地基模型
two-shot method 双液法(化学灌浆)
two-wall sheet-piling cofferdam 双排板桩围堰

type of site soil 场地类型

U

ultimate analysis 极限分析
ultimate balance theory 极限平衡理论
ultimate bearing capacity 极限承载力
ultimate bearing capacity of subsoil
　　　地基极限承载力
ultimate bearing resistance 极限承载阻抗
ultimate bending moment 极限弯曲力矩
ultimate bending strength 极限弯曲强度
ultimate bond strength 极限粘结强度
ultimate breaking load 极限破坏荷载
ultimate carrying capacity of single pile
　　　单桩极限承载力
ultimate compression(compressive) strength
　　　极限抗压强度
ultimate compressive strain 极限压应变
ultimate flexural capacity 极限弯曲能力
ultimate friction angle 极限摩擦角
ultimate lateral resistance of single pile
　　　单桩横向极限承载力
ultimate limit state 极限状态
ultimate load 极限荷载，最大荷载
ultimate load design method
　　　极限荷载设计法
ultimate load method 极限荷载法
ultimate load of pile 桩的极限荷载
ultimate load of subsoil 地基土极限荷载
ultimate pressure 极限压力
ultimate pullout capacity 极限抗拔力
ultimate settlement 最终沉降量

ultimate shearing strength 极限抗剪强度
ultimate strength 极限强度
ultimate strength design 极限强度设计
ultimate strength value 强度终值
ultimate stress 极限应力
ultimate stressed state 极限应力状态
ultimate tensile strain 极限拉应变
ultimate tensile strength 极限抗拉强度
ultimate value 极限值
ultra fine grain 超细颗粒
ultrasonic extractor 超声波抗拔机
ultrasonic wave test 超声波试验
unaltered rock 未变质岩石
unbalanced load 不平衡荷载
unbalanced moments 不平衡力矩
unbalanced thrust transmission method
　　不平衡推力传递法
unbraced excavation 无支撑挖掘
uncased pile 无套管桩
uncombined water 未结合水
unconfined compression strength 无侧限抗压强度
unconfined compression test 无侧限压缩试验
unconfined compressive strength test
　　无侧限抗压强度试验
unconfined groundwater 非承压地下水
unconfined water body 非承压水体
unconnected course 分离层
unconsolidated—drained test
　　不固结排水剪切试验
unconsolidated—undrained triaxial compression strength 不排水三轴剪切强度
unconsolidated-undrained triaxial test
　　不固结不排水试验(UU)
uncovering 剥离

uncovering of foundation 基础露出
under reamed bored pile 扩底钻孔桩
under reamed foundation 扩底基础
under reaming 扩底
underbalanced drilling 负压钻井
underconsolidated soil 欠固结土
underconsolidation 欠固结
underconsolidation clay 欠固结黏土，欠压密黏土
undercut 暗掘，潜挖
underground 地下
underground blasting 地下爆破
underground chamber 地下室
underground diaphragm wall 地下连续墙
underground engineering 地下工程
underground erosion 地下潜蚀
underground excavation 地下开挖
underground exploration 地下勘探
underground grouting works 地下灌浆工程
underground leakage 地下渗流
underground opening 地下硐室
underground rupture plane 深部破裂面
underground store for explosives 地下炸药库
underground tunnel 地下隧道
underground water 地下水，潜水
underground water level 地下水位
underground work 地下工程
underlaying bed 垫底，基础
underlying soil 下卧土
underlying stratum 下卧层
underpinned pile 托换桩
underpinning 托换技术，基础托换
underpinning method 基础托换法
underpinning technique 托换技术
under-reamed bored pile of large diameter

大直径扩孔灌注桩
under-reamed bored pile 钻孔扩底灌注桩
under-reamed foundation 扩底基础
under-reamed pile 扩底桩
undersaturation 不饱和
undersea tunnel 海底隧道
underseepage 地下渗流
underwater blasting 水下爆破
underwater pile driving 水下打桩
underwater tunnel 水底隧道
underwater work 水下工程
undesirable geologic phenomena 不良地质现象
undisturbed ground 未扰动土地
undisturbed soil sample 不扰动土样，原状土样
undisturbed soil sampling 取原状土样
undrained loading 不排水加载
undrained settlement 瞬时沉降
undrained shear strength 不排水抗剪强度
undrained shear test 不排水剪切试验
undrained triaxial test 不排水三轴试验
unequal(uneven)settlement 不均匀沉降
unevenly compressible foundation
　　　　不均匀压缩地基
unfilled caisson foundation 空心沉井基础
unfilled porosity 非饱和孔隙度
unfrozened water content 未冻水量
uniaxial compression 单轴向压缩
uniaxial compression force 单轴压力
uniaxial compression strain of rock
　　　　岩石单轴压缩应变
uniaxial compression strength of rock
　　　　岩石单轴压缩强度
uniaxial compression test 单轴压缩试验
uniaxial stress 单轴拉力

uniaxial tensile stress 单轴抗拉强度
uniaxial tension strength test 单轴抗拉强度试验
unidimensional consolidation 单向固结，一维固结
unified soil classification system 统一土壤分类法
uniform compaction 均匀压实
uniform grading 均匀级配
uniform settlement 均匀沉降
uniform size aggregate 均匀粒径骨料
uniform soil 级配均匀的土
uniform strength 等强度
uniform stress 均布应力
uniformity coefficient 均匀系数
uniformly distributed load 均布荷载
uniformly graded 均匀级配
unit area 单位面积
unit area loading 单位面积荷载
unit deformation 单位变形
unit load 单位荷载
unit normal vector 单位法向矢量
unit shaft resistance 单位侧阻
unit skin friction 单位表面摩擦力
unit strain 单位应变
unit stress 单位应力
unit tamping energy 单位夯击能量
unit trench section 单元槽段
unit volume expansion 单位体积膨胀
unit weight 容重，表观密度，重度
unloading 卸载
unloading curve 卸载曲线
unloading modulus 卸载模量
unloading test 卸荷试验
unsafe foundation depth 不安全的基础埋置深度
unsaturated flow 非饱和流
unsaturated soil 非饱和土

unsaturation test 非饱和试验
unslacked lime 生石灰
unslacked lime pile 生石灰桩
unstable slope 不稳定坡
unstable state 不稳定状态
unstable stratification
　　　　不稳定层理，不稳定分层
unsteady flow 非稳定流
unsteady seepage 非稳定渗流
unstratified soil 非层状土
unsupported height 自由高度
unsymmetrical footing 不对称基础
unsymmetrical load 不对称荷载
unwatering 疏干
unweathered rock 新鲜岩石
upcast shaft 回风井
upheave 隆起，抬起
uphill 上坡
up-hole method 上孔法
uplift force 浮力
uplift load test of pile 桩的抗拔试验
uplift pile 抗拔桩
uplift pressure 扬压力
uplift resistance 抗拔力
upper-lower limit of slope protection
　　　　护坡的上下限
upsetting moment 倾覆力矩
upstream cofferdam 上游围堰
U-tube manometer U形管测压计

V

vacuum method 真空预压法

vacuum preload(ing)　真空预压
vacuum preloading method　真空预压法
vacuum pump　真空泵
vacuum sampling tubes　真空取样器
vacuum triaxial test　真空三轴试验
vacuum well point　真空井点
vacuum well system　真空井点系统
vadose water　上层滞水，渗流水，土壤水
vadose zone　包气带，渗流区
valley flat　河漫滩
valley terrace　河谷阶地
Van der Waals force　范德华力
Vane borer　十字板仪
vane penetrometer　十字板贯入仪
vane shear apparatus　十字板剪力仪
vane shear test　十字板剪切试验
vane strength　十字板抗剪强度
variable frequency vibrator　变频振动器
variable head permeability test
　　　　变水头渗透试验
variable module model　变模量模型
varved clay　带状黏土，分层黏土
vegetation cover of slope protection　植树护坡
vein　矿脉
velocity of permeability　渗透速度
ventilation lateral　通风平巷
vertical allowable load capacity of single pile
　　　　单桩竖向抗压极限承载力
vertical excavation method　竖向开挖法
vertical loading test of pile
　　　　桩的荷载试验，桩的抗压试验
vertical pile　竖直桩
vertical shaft　竖井，立井
vertical shrinkage　竖向收缩率

vertical stiffness factor 地基抗压刚度系数，地基竖向刚度系数
vertical ultimate carrying capacity of single pile 单桩竖向抗压极限承载力
vertical ultimate load capacity of pile groups 群桩竖向极限承载力
vertical ultimate uplift resistance of single pile 单桩竖向抗拔极限承载力
Vesic's ultimate bearing capacity formula 魏锡克极限承载力公式
viaduct 高架道路，高架桥
vibrating base plate compactor 振动平板压实机
vibrating compactor 振动压实机
vibrating extractor 振动拔桩机
vibrating load 振动荷载
vibrating pile hammer 振动打锤机
vibrating plate compactor 板式振动压实机
vibrating wire cell 钢弦式压力盒
vibrating wire strain gauge 钢弦式应变仪
vibration compaction 振动法压实
vibration isolation 隔振
vibration isolation of foundation 基础隔震
vibration pile-driver 振动打桩机
vibration replacement stone column 振冲碎石桩
vibration screen 振动筛
vibration triaxial apparatus 振动三轴仪
vibration-proof foundation 防震基础
vibrator 振动器
vibrator sunk pile 振沉桩
vibratory bored pile 干振成孔灌注桩
vibratory compaction 振动压实
vibratory driver 振动打桩机
vibratory liquidation test 振动液化试验
vibratory pile hammer 振动桩锤

vibro drilling 振动钻井
vibro driver 振动打桩机
vibro tamper 振动夯
vibro-compaction method 振冲密实法
vibrocorer 振动取芯器
vibrocoring method 振动取样方法
vibro-densification method 振密挤密法
vibro-driver extractor 振动拔桩机
vibroflot 振冲器
vibroflotation method 振冲法
vibro-pile 振动灌注桩
vibro-pile driver 振动打桩机
vibro-rammer 振动夯
vibro-replacement 振动置换
vibro-replacement method 振冲置换法
vibro-replacement stone column 振冲碎石桩
vibro-roller 振动碾
vibrosinking method 振动沉桩法
vibrosinking of pile 振动沉桩
virgin compression curve 原始压缩曲线
virgin stress 原始应力
virginal overburden pressure 原始上覆压力
virtual stress 虚应力
visco slip fault 黏滑断层
viscoelastic behavior 黏弹性
viscoelastic foundation 黏弹性地基
viscoelastic model 黏弹性模型
viscoelastic property 黏弹特性
viscoelasticity 黏弹性
viscoplastic fluid 黏塑性流体
viscosity 黏滞性
viscous damping 黏滞阻尼
viscous flow 黏滞流
visual soil classification 土的肉眼分类

void　空眼
void content　孔隙含量
void measurement apparatus　孔隙测定仪
void ratio　孔隙比，孔隙率
void ratio in densest state　最小孔隙比
void ratio in loosest state　最大孔隙比
void water　孔隙水
volcanic ash　火山灰
volcanic earthquake　火山地震
volcanic rock　火山岩
volcanic saprolite　火山岩风化土
volume change　体胀率
volume expansion　体胀率
volume injected　灌入量
volume of earthwork　土方工程量
volume shrinkage ratio　体缩率
volumetric coefficient　体积系数
volumetric deformation modulus　体积变形模量
volumetric flask　比重瓶
volumetric heat capacity　体积热容量
volumetric shrinkage　体缩率
volumetric specific heat　体积比热
volumetric strain　体应变
V-shape cut　楔形掏槽

W

wagon drill　汽车钻机
waist(ing)　缩颈
wales　撑梁
wall adhesion　墙黏着力
wall foundation　墙基，墙式基础
wall friction　墙摩擦力

wall of a well 井壁
wall panel 槽段
wall protection by slurry 泥浆护壁
wall rock 围岩
wall stabilization with clear water 清水护壁
wall-protecting slurry 护壁泥浆
warp clay 淤积黏土
wash 冲积物，冲刷
wash boring 冲洗钻孔
wash out 水冲刷
wash pipe 冲洗管
wash point penetrometer 水冲式贯入仪
wash sample 冲洗样
washout 冲蚀
waste 弃土
waste containment 废料污染
waste disposal 废料处理
waste disposal facilities 废料处理设备
waste geotechnics 废料岩土力学
waste mantle 废料覆盖层
water bearing capacity 容水量
water bearing bed(layer) 含水层
water cement ratio 水灰比
water cement slurry 水泥浆
water conservancy project 水利工程
water content 含水率
water content ratio 含水比
water content test 含水率试验
water course 水流路径
water creep 渗水
water displacement method 换水法(测密度)
water film theory 水膜理论
water flush boring 水冲钻探
water glass 水玻璃，硅酸钠

water glass solution 水玻璃溶液
water holding capacity 保水能力
water holding rate 保水率
water in soil 土中水
water inflow 涌水量
water injecting test 注水试验
water intake tunnel 引水隧洞
water jet driver 射水打桩机
water level gauge 水位计
water level survey 水准测量
water logged soil 浸湿土
water of crystallization 结晶水
water of infiltration(percolation)
　　　　　渗入水，渗透水
water permeability test 透水性试验
water plane 潜水面
water pressure test 压水试验
water pressure test hole 压水试验孔
water pumping test 抽水试验
water pumping test hole 抽水试验孔
water quality analysis 水质分析
water replacement method 换水法(测密度)
water retaining capacity 持水度，饱水能力
water sample 水样
water sampler 取水样器
water sealing 水封
water sensitive soil 湿陷性土
water stop plate for lining 衬砌止水板
water stopping 止水
water stopping band 止水带
water stopping rubber gasket 止水胶垫
water supply tunnel 给水隧道
water table 地下水位
water test 渗水测试

water test log 压水试验记录
water tight 不透水
water tight core 防水心墙，不透水心墙
water tight screen 止水帷幕
water treatment 水处理
water tunnel 水工隧道
water void ratio 水隙比
water-bearing stratum 含水层
water-fed rock drilling 湿式凿岩机
waterlogging 积水现象
water-plasticity ratio 液性指数
waterproof charge 防水药包
waterproof concrete 防水混凝土
waterproofing of tunnel 隧道防水
watershed tunnel 越岭隧道
waterstop blade 止水片
watertight 防水，防渗的，不透水的
watertight joint 防水接缝
watertight screen 止水帷幕
watertightness 防水性，不透水性
wave built terrace 浪成阶地
wave cutting 浪蚀
wave equation analysis 波动方程分析
wave velocity method 波速法
weak foundation(ground) 软弱地基
weak intercalated layer 软弱夹层
weak rock 软岩
weak soil 软土
weak structural plane 软弱结构面
weak zone 软带
weakly weathered layer 弱风化层
weakly weathered zone 弱风化带
weatherability 耐风化性
weathered crust 风化壳

weathered granite 风化花岗石
weathered layer 风化层
weathered rock 风化岩石
weathered rock-soil 风化料
weathered zone 风化带
weathering 风化作用，风化
weathering crust 风化壳
weathering degree of rock 岩石风化程度
weathering zone 风化带
wedge cut 楔形掏槽
wedge failure analysis of rock slope
　　　　岩坡楔体破坏分析
wedge theory 楔体理论
wedging effect 楔入效应
weep hole 泄水孔
Wei Rulong-Khosla-Wu model
　　　　魏汝龙-Khosla-Wu 模型
weight density of soil 土的重度
weight in water method 水中称重法（测密度）
weight sounding 加载触探仪
weighted mean value 加权平均值
weir 堰
well 井
well drain 排水井
well filter 滤井
well foundation 井筒基础
well point 井点排水
well point dewatering 井点降水
well point method 井点法排水
well point pumping system 井点排水系统
well point system 井点系统（轻型）
well pumping test 抽水试验
well resistance 井阻
well screen 井滤网

well-graded soil　　良好级配土
wet analysis　　湿分析法
wet compaction　　偏湿压实
wet density　　湿密度
wet material　　湿土
wet mechanical analysis　　湿法粒径分析
wet sample　　湿试样
wet screening　　湿法筛分
wet swelling　　湿胀
wet unit weight　　湿容重
wet-mix spraying process　　湿式喷射法
wetness index　　湿度指数
wet-subsidence due to overburden
　　　　计算自重湿陷量
wet-subsidence for classification　　分级湿陷量
wetting and drying test　　干湿试验
wetting front　　浸润线
wick drain　　袋装砂井
wind abrasion　　风蚀
wind drift sand　　风砂，流砂
wind laid deposit　　风沉积
Winkler foundation　　文克勒地基
Winkler foundation model　　文克尔地基模型
Winkler's assumption　　文克勒假定
wire mesh-reinforced shotcrete
　　　　钢丝网加固喷射混凝土
wooden sheet pile　　木板桩
work hardening　　加工硬化
work softening　　加工软化
work stone　　料石
workability　　和易性
working chamber of caisson　　沉箱工作室
working level of piling rig　　打桩钻机工作面
working load　　使用荷载，工作荷载

working pile 工作桩
working platform 工作平台
worksite 工地
woven fabrics(geotextile)
　　　　纺成土工织物，有纺布

Y

yield 屈服
yield criteria 屈服准则
yield function 屈服函数
yield locus 屈服轨迹
yield of foundation 基础沉陷
yield of water 出水量
yield of well 井出水量
yield point 屈服点
yield strength 屈服强度
yield stress 屈服应力
yield stress model 屈服应力模型
yield surface 屈服面
yield value 屈服值
yielding soil 流动土，软土
young clay 新近沉积黏土
young loess 新黄土
Young's modulus 弹性模量

Z

zero air void ratio 饱和孔隙比
zero air voids density 饱和密度
zero air void curve 饱和曲线
zeolite 沸石

Zemochkin's method 热摩奇金法
zero isochrone 零时水坡线
zone of capillarity 毛细带
zone of capillary saturation 毛细管水饱和带
zone of differential settlement 不均匀沉陷区
zone of saturation 饱和带
zonal soil 区域性土
zone of plastic flow 塑流区
zero lateral strain test 无侧向应变试验

汉英部分

A

阿米水云母　ammersooite
阿太堡界限　Atterberg limits
埃(Å)　angstrom
埃洛石　halloysite
艾德测头　Idel sonde
安全承载能力　safe bearing capacity, safety load carrying capacity
安全储备　safe margin
安全措施　safe measure
安全地层　safe formation
安全度　level of safety
安全分析　safety analysis
安全距离　safety distance, separation distance
安全水泥　sound cement
安全系数　factor of safety, safe factor, safety coefficient(factor)
安全钻井　safe drilling
安山凝灰岩　andesite-tuff
安山玄武岩　andesite basalt
安山岩　andesite
鞍形壳　saddle-shaped shell
铵质土　ammonium soil
岸壁　bulkhead(wall)
岸壁基础　base of a quay wall
岸边冲刷　shore cutting
岸边淤积　shore deposit
岸积物　bank materials
岸砾石　bank gravel

岸上桩　onshore piles
暗管　closed conduit
暗掘　undercut
暗色岩　trap rock
凹坑　pothole
奥陶纪　Ordovician period

B

八面体层　octahedral layer
八面体剪应力　octahedral shearing stress
八面体正应力　octahedral normal stress
巴林石　Barlin stone
巴隆固结理论　Barron's consolidation theory
拔出荷载　pull-out load
拔力　pull out force
拔桩　pile extraction, pile pulling
拔桩机　pile drawer, pile extractor
拔桩力　extractor force
拔桩阻力　pile extraction resistance
坝顶超高　camber
坝机截水墙槽　key trench
坝基　dam foundation
坝脚抛石　talus
坝失事，溃坝　dam failure
坝址测量　dam site survey
坝址勘查　dam site investigation
坝趾　dam toe
坝踵　dam heel
坝座变形　abutment deformation
白垩　chalk
白垩纪　Cretaceous period
白垩泥灰岩　chalk marl

白垩黏土　chalky clay
白云母　muscovite
白云母花岗石　muscovite-granite
白云大理石　dolomite marble
白云石灰石　dolomite limestone
白云岩　dolomite
摆式打桩机　pendulum hammer, pile-driver
斑岩　porphyry
板式承台　pile cap as plate
板式挡土结构　sheet retaining structure
板式基础　slab foundation
板式水平位移计　horizontal plate gauge
板式振动压实机　vibrating plate compactor
板岩　cleaving stone, plate stone, slate
板页岩　plate shale
板桩　sheet pile, sheeting pile
板桩岸墙　sheet-pile bulkhead
板桩堤　sheet-pile levee
板桩防渗墙　sheet pile cut-off wall
板桩加固法　pile slab method, soil stabilization method by sheet pile
板桩结构　sheet pile structure
板桩截水墙　sheet-pile cut-off wall
板桩锚固　sheet pile anchorage
板桩墙　sheet pile wall, sheeting, sheet-pile wall
板桩式挡土墙　sheet-pile retaining wall
板桩围护　sheet pile-braced cuts
板桩围堰　sheeting/sheet-pile cofferdam, sheet-pile enclosure
板桩帷幕　sheet-pile curtain
板桩支撑　sheet piling support
半饱和　semi-saturation
半盾构　semi-shield
半干燥　partial desiccation

半机械化盾构	semi-mechanized shield
半坚硬岩石	semi-hard rock
半经验公式	semi-empirical formula
半空间	half-space
半逆解法	semi-inverse method
半深海沉积	bathyal deposit, moderate sea deposit
半无限弹性体	semi-infinite elastic body
半无限平面问题	semi-infinite plane problem
半无限体	semi-infinite solid
半咸沉积	brackish deposit
半咸水	brackish water
拌合加固法	admixture stabilization, stabilization method by mixture
拌合桩	mixed-in-place pile
拌浆设备	slurry equipment
拌砂	sand cutting
包气带	vadose zone
包气带水	aeration zone water
饱和	saturation
饱和百分率	saturation percentage
饱和比	saturation ratio
饱和比湿度	saturation specific humidity
饱和参数	parameter of saturation
饱和测试	saturation testing
饱和层	saturated zone
饱和差	saturation deficiency(deficit)
饱和场	saturation field
饱和程度试验	saturation test
饱和带	saturated belt, zone of saturation
饱和的	saturated
饱和点	saturation point
饱和度	degree of saturation, percent saturation, saturation degree

饱和分析法	saturation analysis
饱和含湿量	saturation humidity ratio
饱和含水量	saturation capacity
饱和含水率	percentage of saturated water content, saturation moisture content
饱和孔隙比	zero air void ratio
饱和力	saturation force
饱和密度	saturated density, zero air voids density
饱和面	plane of saturation, saturated surface
饱和浓度	saturation concentration
饱和区	saturation zone
饱和区监测井	saturated zone monitoring well
饱和曲线	saturation curve, zero air void curve, saturation line
饱和溶液	saturated solution
饱和砂	saturated sand
饱和湿胀应力	saturation swelling stress
饱和时间	saturation time
饱和水	saturated water
饱和梯度	saturation gradient
饱和透水性	saturated permeability
饱和土	saturated soil
饱和土密度	density of saturated soil
饱和温度	saturation temperature
饱和吸水率	coefficient of water saturation
饱和系数	saturation coefficient
饱和线	line of saturation
饱和压力	saturation pressure
饱和液	saturated liquid
饱和蒸汽	saturated steam, saturation steam
饱和蒸汽温度	saturation steam temperature
饱和指数	saturation index
饱和重度	saturated unit weight

饱和状态	saturated state, saturation condition, state of saturation
饱气水	aerated water
饱水能力	water retaining capacity
保留强度	reserved strength
保湿室	moisture room
保守设计	overdesign
保水率	water holding rate
保水能力	water holding capacity
保水性	moisture retention
爆扩桩	blasting-expand pile, blown tip pile, exploded pile
爆破	shooting, shot, blasting
爆破的岩石	shot rock
爆破地震危险半径	radius of seismic danger due to blasting
爆破加固	blasting consolidation
爆破振密	blast densification
爆炸法	dynamite method
爆炸挤密	explosive compaction
爆炸挤密法	blasting compaction method, densification by explosion
杯口基础	socket foundation
北达科他州圆锥试验	North Dakota cone test
北极冰沼土	tundra soils in the arctic area
背拱	rear arch
背斜	anticline
钡长石	Baryta feldspar, celsian feldspar
被动朗肯土压力	passive Rankine pressure
被动破坏	passive failure
被动塑性平衡状态	passive state of plastic equilibrium
被动土压力	passive earth pressure
被动土压力系数	coefficient of passive earth

pressure
被动桩　passive pile
本构定律　constitutive law
本构方程　constitutive equations
本构关系　constitutive relation
本构微分方程　differential equation of constitution
崩滑　slumping
崩积层　colluvium
崩积土　colluvial clay(soil)
崩积土图样　sketch for the colluvial soils
崩解性　disintegrative
崩解岩石　crumbling rock
崩落　dilapidation, scree
崩落地带　slump belt
崩落线　breaking edge
崩塌　devolution, slacktip
崩塌坝　avalanche dam
崩塌土　collapsing soil
比奥固结理论　Biot consolidation theory
比表面　relative surface area
比表面积　specific surface area
比表面积率　rate of specific surface area
比表面积试验　specific surface test
比表面能　specific surface energy
比表面张力　specific surface tension
比贯入阻力　specific penetration resistance
比例极限　proportional limit
比例加载　proportional loading
比能　specific energy
比黏度　specific viscosity
比强度　specific strength
比容　specific volume
比吸浆量　specific grout absorption

比吸水量 specific absorption
比徐变 specific creep
比重计 areometer
比重计分析 hydrometer analysis
比重瓶 density bottle, picnometer, pycnometer, volumetric flask
比重试验 specific gravity test
比阻尼容量 specific gravity capacity
毕肖普简化条分法 Bishop's simplified method of slices
毕肖普与摩根斯坦法 Bishop and Morgenstern slope stability analysis method
碧玄岩 basanite
边对面絮凝 edge-to-face flocculation
边墩压力 side abutment pressure
边沟 side drain
边界层 boundary layer
边界角 rim angle
边界条件 boundary conditions
边界余量 residue of boundary
边界桩 border pile
边跨 side span
边坡 side slope
边坡安全系数 safety factor of slope
边坡安全性标准 strength criterion of slope safety
边坡安全性的位移判据 displacement criterion of slop safety
边坡剥落 slope flaking
边坡长期稳定 long term stability of slope
边坡冲刷 slope erosion
边坡反分析 slope inversion analysis
边坡加固 slope stabilization

边坡监测　slope monitoring
边坡坡度　side slope grade
边坡坡度容许值　allowable grade of slope
边坡瞬时稳定　instantaneous stability of slope
边坡稳定安全系数　safety factor of stability of slope
边坡系数　side slope coefficient
边坡桩　slope stake
边缘扰动　edge disturbance
边缘压力　edge pressure
边桩　side piling
编织土工布　knitted geotextile
扁拱　shallow arch
扁平度　flatness ratio
扁千斤顶　flat jack
扁千斤顶法　flat jack technique
扁式松胀仪　flat dilatometer
扁式液压顶法　the method of flat jack
便道　access road
变截面　reduce section
变模量模型　variable module model
变频振动器　variable frequency vibrator
变色黏土　discoloured clay
变水头渗透试验　falling head permeability test, variable head permeability test
变水头渗透仪　falling head permeameter
变形　deformation
变形参数试验　deformation parameters test
变形分辨率　deformation resolution
变形控制　deformation control
变形力　strain forces
变形模量　modulus of deformation
变形条件　deformation conditions
变形仪　deformeter

变形应力　deformation stress
变质岩　metamorphic rock
变质岩石学　metamorphic petrology
标定　calibration
标定曲线　calibration curve
标定试验　calibration test
标高　elevation
标准层　key bed
标准冻结深度　standard frost depth
标准贯入击数　blow count of SPT
标准贯入试验　standard penetration test(SPT)
标准化　normalization
标准化石　guide fossil
标准砂　standard sand
标准筛　standard screen(sieve)
表层地基加固法　shallow ground stabilization
表层冻土　shallow freezing
表层土　topsoil
表层压密法　surface compaction
表层应力　skin stress
表观比重　apparent specific gravity
表观抗剪强度　apparent shearing strength
表观流速　apparent velocity
表观摩擦系数　apparent coefficient of friction
表观内摩擦角　apparent angle of internal friction
表观黏度　apparent viscosity
表观黏聚力　apparent cohesion
表观先期固结压力　apparent preconsolidation pressure
表流　surface flow
表面波法　surface wave velocity method
表面波试验　surface wave test
表面沉降板　surface settlement plate

表面处理　resurfacing
表面处理　surface treatment
表面粗糙度　roughness of the surface
表面力　surface force
表面摩擦　skin friction, surface friction
表面摩擦力　skin-friction force
表面排水　face drain
表面纹理　surface texture
表面压实　superficial compaction
表面硬化　case hardening
表面张力　surface tension
表土　surficial soil
滨(宾)汉模型　Bingham model
滨海沉积　littoral deposit
滨海带　littoral zone
滨汉流体　Bingham fluid
滨汉体　Bingham substance
滨后阶地　backshore terrace
冰长石　adularia
冰成岩　glacial rock
冰川　glacier
冰川边缘沉积，岸边淤积　inwash
冰川沉积　glacial deposit
冰川沉积　ice-laid deposits
冰川沉积平原　outwash plain
冰川黄土　glacial loess
冰川角砾　glacial breccia
冰川碎石带　dirt band
冰川学　glaciology
冰点　freezing point
冰冻线　frost line
冰碛　moraine
冰碛沉积　morainal deposit
冰碛堆积物　drift

冰碛黏土　glacial clay
冰碛平原　apron plain
冰碛土　drifted soil, glacial till, till
冰蚀　glacial erosion
冰水沉积　aqueoglacial deposit, fluvio-glacial deposit, glacial outwash
冰水沉积土　glacial-fluvial soils
冰透镜体　ice lenses
冰压力　ice pressure
丙凝灌浆　acrylamide grouting
波动方程分析　wave equation analysis
波罗杆(沉降观测)　Borros point
波美比重　Baume gravity
波美比重计　Baume hydrometer
波美比重标　Baume scale
波美度　degree Baume
波士顿蓝黏土　Boston blue clay
波速法　wave velocity method
玻璃量杯　measuring glass
剥离　uncovering
剥离比　rate of stripping
剥蚀　degradation
剥蚀台地　abrasion table-land
剥蚀作用　denudation
伯格模型　Burger model
泊(黏度单位)　poise, P
泊松比　Poisson's ratio
薄板理论　plate theory
薄壁取土器　thin wall sampler
薄层构造　lamella structure
薄层黏土　paper clay
薄膜法　membrane method
薄膜式压力盒　membrane type pressure gauges
薄膜水　film water

补偿切线模量　offset tangent modulus
补偿式基础　compensated foundation
补充勘测　additional survey
补充孔　additional borehole
补给　alimentation
补给率　recharge rate
补给区　alimentation area, recharge area
补救　remedy
补救灌浆,补强灌浆　remedial grouting
补强　strengthening
补强层　strengthen layer
补强处理　remedial treatment
补强灌浆　strengthening grouting
不安全的基础埋置深度　unsafe foundation depth
不饱和　undersaturation
不变量　invariant
不对称荷载　unsymmetrical load
不对称基础　unsymmetrical footing
不固结不排水试验(UU)　unconsolidated-undrained triaxial test
不固结排水剪切试验　unconsolidated drained test
不规则变形　irregular deformation
不规则层理　irregular bedding
不规则荷载　irregular load
不规则颗粒形状　irregular particle shape
不规则裂缝　random crack
不规则曲线　irregular curve
不规则土层剖面　erratic soil profile
不规则岩石试件　irregular specimen for rocks
不均匀沉降　nonuniform settlement, unequal settlement, uneven settlement
不均匀沉陷区　zone of differential settlement
不均匀冻胀　differential frost heave

不均匀级配　nonuniform grading
不均匀隆胀　differential heave
不均匀系数　coefficient of uniformity
不均匀性　nonhomogeneity
不均匀压缩地基　unevenly compressible foundation
不可压缩性　incompressibility
不连续构造面　discontinuity structural plane
不连续滑动面　broken sliding surface
不连续级配　gap gradation
不连续级配曲线　skip grading curve
不连续级配土　gap-graded soil
不良地质条件　difficult ground condition
不良地质现象　adverse geologic phenomena, undesirable geologic phenomena
不良级配土　poorly-graded soil
不排水加载　undrained loading
不排水剪切试验　undrained shear test
不排水抗剪强度　undrained shear strength
不排水三轴剪切强度　unconsolidated undrained triaxial compression strength
不排水三轴试验　undrained triaxial test
不排水下沉法　method of undrained sinking
不平衡荷载　unbalanced load
不平衡力矩　unbalanced moments
不平衡推力传递法　unbalanced thrust transmission method
不扰动土样　undisturbed soil sample
不透水　water tight
不透水边界　impervious boundary
不透水层　aquifuge, impermeable barrier, impervious layer

不透水衬层　impervious liner
不透水的　impervious
不透水的坝心墙　core wall of dam
不透水地层　impervious stratum
不透水地基　impervious foundation
不透水系数　coefficient of imperviousness
不透水心墙　impervious core
不完全饱和　partial saturation
不完全固结　incomplete consolidation
不完全渗水井　partial penetrating well
不稳定层理，不稳定分层　unstable stratification
不稳定坡　unstable slope
不稳定性　instability
不稳定指数　instability index
不稳定状态　unstable state
布尔登压力计　Bourdon gauge
布莱恩细度　Blaine fineness
布朗运动　Brownian movement
布辛涅斯克理论　Boussinesq theory
部分安全系数　partial safety factor
部分饱和土　partially saturated soil
部分风化带　partially weathered zone
部分浸水　partial submergence
部分排水加载　partially drained loading

C

擦痕　striation
擦痕卵石　scraping pebble, striated pebble
擦痕面　slickenside, striated rock surface
材料非线性　material non-linearity
材料非线性问题　nonlinear material problem
材料强度系数　strength factor of material

材料特性　material properties
材料阻尼　material damping
采空塌陷　mining subsidence
采料场　borrow area
采石场　quarry
采石坑砾石　bank-run gravel
采芯器　core catcher
采样　sampling
参数　parameter
参数辨识　parameter identification
参数反分析　inverse analysis of parameters
参数反演　parameter inversion
参数方程　parameter equation
参数估计　parameter estimation
参数检测　parameter detection
参数曲线　parameter curve
参数优化　parameter optimization
残差分析　residual analysis
残积层　eluvium
残积土　eluvial soil, residual soil
残积相　eluvial facies
残留层理　relict bedding
残坡积膨胀土　residual diluvial expansive soil
残岩　remaining rock
残余饱和度　residual saturation
残余变形　residual deformation
残余沉积　residual deposite
残余沉降　residual settlement
残余结构　relic structure
残余抗剪强度　residual shear strength
残余抗拉强度　residual tensile strength
残余孔隙水压力　residual pore water pressure
残余内摩擦角　residual angle of internal friction

残余黏聚力　residual cohesion intercept
残余膨胀　permanent expansion
残余强度　residual strength
残余挠度　residual deflection
残余收缩　residual shrinkage
残余水饱和度　irreducible water saturation
残余水分　residual moisture
残余塑流　after effect
残余土　saprolith
残余应变　residual strain
残余应变场　residual strain field
残余应力　remaining stress, residual stress
残余应力场　residual stress field
残值　residual value
仓效应　bin effect
操作规程　operating rules
槽壁稳定性验算　stability analysis of diaphragm trenches
槽段　wall panel
槽孔支撑　shoring of trench
槽探　trenching
草皮护坡　sod revetment, turf protection of slope
草图　sketch
草土围堰　straw and earth cofferdam
侧壁导坑法　side drift method
侧壁钻机　sidewall drill
侧边弯曲　side camber
侧沟　side ditch
侧力　side force
侧立面　side elevation
侧面加载　side loading
侧面摩擦　side friction
侧面掏槽　sidewise cut
侧扭屈曲　lateral torsional buckling

侧弯试件	side bend specimen
侧弯试验	side bend test
侧限加固	lateral restraint reinforcement
侧限抗压强度	confined compressive strength
侧限压力	lateral confining pressure
侧限压缩模量	oedometric modulus
侧限压缩试验	confined compression test
侧向变形	lateral deformation
侧向冲刷	lateral scouring
侧向地基反力系数	lateral modulus of subgrade reaction
侧向负荷	side load
侧向刚度	lateral rigidity
侧向滑动	lateral slide
侧向滑移	side skid
侧向挤压	lateral compression
侧向开挖	lateral cutting, side cutting, side excavation
侧向抗滑力	sideway skid resistance
侧向扩展	lateral spread
侧向排水能力	lateral drainage ability
侧向膨胀	lateral expansion
侧向屈服	lateral yield
侧向收缩	lateral shrinkage
侧向土压力	lateral earth pressure
侧向托换	lateral underpinning
侧向位移	lateral displacement
侧向稳定性	lateral stability
侧向压屈	lateral buckling
侧向移动标桩	lateral movement stake
侧向应变	lateral strain
侧向应变指示器	lateral strain indicator
侧向应力	lateral stress
侧向约束	lateral restraint

侧压力　lateral pressure, side compression, side pressure
侧压力系数　lateral pressure coefficient
侧压系数　coefficient of horizontal pressure
侧压仪　lateral pressure apparatus
侧移刚度　sidesway stiffness
侧移弯矩　sidesway moment
侧钻水平井　sidetrack horizontal well
测定地质图　measured geological map
测缝计　crack gauge, joint gauge, joint meter
测径记录　caliper log
测孔　sampling hole
测力计　dynamometer, load gauge
测深仪　deep sounding apparatus
测渗计　lysimeter
测微计　micrometer
测斜仪　clinometer, inclinometer, deflection inclinometer, slope level
测压管　piezometer tube, pressure measurement probe
测压管水面　piezometric surface
测压管水头　piezometric head
测压管水位　piezometric level, potentiometer surface
测压计　pressure gage
测压孔　piezometer opening, pressure hole, pressure tapping hole
测压状态　piezometric regime
层间不冻层　talik
层间冻土　intergelisol
层间附着力　ply adhesion
层间含水层　confined aquifer
层间滑动　interformational sliding
层间滑动断裂　interlayer-gliding fault

层间剪切强度　interlaminar shear strength
层间接触　interlayer contact
层间距　interlayer spacing
层间连续性　interlayer continuity
层间联结　interlayer bonding
层间摩擦　interfacial friction
层间软弱面强度　shear strength of interlayered weak surface
层间水　interlayer water, intermediate water, middle water, interstrated water
层间系数　coefficient between layers
层理　bedding
层理面　stratification plane
层流　laminar flow
层面　bedding plane
层面滑动　bedding slip
层状地层　bedded formation, layered strata
层状地基　stratification foundation
层状地貌　stratified relief
层状构造　bedded structure
层状结构　laminated structure, layer structure, stratified structure
层状介质　layered media
层状黏土　laminated clay, layered clay
层状水　stratum water
层状体　lamina
层状体系　layered system
层状土　laminated soil, stratified soil
插板　spile
插桩　pitching of pile
查勘　perambulation
查勘测量　reconnaissance survey
差动式岩石锚杆引伸仪　differential rock bolt extensometer

差热分析	thermal differential analysis
柴排压沉	ballasting of mattress
柴油打桩锤	diesel pile hammer
掺合料	additive
掺合物	admixtures
掺气剂	air-entraining chemical compound
掺气渗透仪	air-entry permeameter
掺砂水泥	sand cement
掺土水泥	soil cement
产状	attitude, occurrence
铲车	shovel car
铲运机	bowl scraper, hauling scraper
长高比	length height ratio
长径比	length to diameter ratio, ratio of length-diameter
长期荷载	long-time loading
长期荷载试验	long term loading test
长期荷载作用下的强度	strength under sustained load
长期模量	long-term modulus
长期蠕变	long-term creep
长期试验	long run test
长期稳定性	long-term stability
长期性状	long-term behavior
长石	feldspar
长石砂岩	arkose, feldspathic sandstone
长石石英砂	feldspathic quartz sandstone
长石石英岩	arkose quartzite
长细比	slender ratio
常规分析	routine analysis
常规固结试验	routine consolidation test
常规观测	routine observation
常规监测	routine monitoring
常水头渗透试验	constant head permeability test

常水头渗透仪 constant head permeameter
常水位 ordinary water level
场地地质适宜性 geological suitability of site
场地复杂性 complexity of site
场地覆盖层厚度 thickness of site soil layer
场地勘查 site exploration
场地类型 type of site soil
场地烈度 site intensity
场地平整 site grading
敞口灌浆 open-ended grouting
超打 overdriving
超打桩 overdriven pile
超弹性定律 hyperelastic law
超固结 overconsolidation
超固结比 over consolidation ratio(OCR)
超固结黏土 overconsolidated clay
超固结土 overconsolidated soil, preloaded soil, overconsolidation soil
超过屈服强度的应力 stress beyond yield strength
超径材料 oversized materials
超径颗粒 oversize particle
超静水压力 excess hydrostatic pressure, hydrostatic excess pressure
超孔隙水压力 excess pore water pressure, u_e
超灵敏黏土 extra-sensitive clay
超前灌浆 advanced grouting
超声波抗拔机 ultrasonic extractor
超声波试验 ultrasonic wave test
超声波验桩法 sonic coring
超填 overfill
超挖 overexcavate, overbreak
超细颗粒 ultra fine grain
超限应变 overstrain

超限应力　overstress
超压实黏土　overcompacted clay
超应力荷载　overstress load
超应力岩体　overstressed rock mass
超载　excess load, overcharge, overload, surcharge
超载等效高度　equivalent height of surcharge
超载能力　overload capacity
超载试验　over load test
超载特性　overload characteristics
超载预压　surcharge preloading
超张拉　over stretching
超重　overweight
超阻尼　overdamping
炒干法　fry-dry method
沉锤法　falling cone method
沉淀　sedimentation
沉淀法粒度分析　sediment analysis
沉管法　immersed tube method
沉管法　pipe sinking
沉管法　sunken tube method
沉管法　tube sinking method
沉积　deposit
沉积的淤泥　sullage
沉积构造　sedimentary tectonics
沉积曲线　sedimentation curve
沉积条件　mode of deposition
沉积土　sedimentary soil
沉积物　sediment
沉积岩　sedimentary rock
沉降　settlement
沉降板　settlement plate
沉降比　settlement ratio
沉降变形监测　monitoring of settlement and

deformation
沉降测量　settlement measurement
沉降测量仪　settlement gauge
沉降差　differential settlement
沉降等值线　settle contour
沉降缝　settlement joint
沉降观测　settlement observation
沉降计算法　method for settlement calculation
沉降计算深度　settlement calculation depth
沉降曲线　settlement curve, settling curve
沉降室　settling chamber
沉降衰减　fading of settlement
沉降速度　settlement velocity, settling rate, settling velocity, sinking speed
沉降速率　rate of settlement
沉降校正系数　settlement correction factor
沉降应力　settlement stress
沉降影响深度　distance of settlement influence
沉降预测　settlement forecast
沉降作用　settlement action
沉井　open caisson, cytinder
沉井壁　open caisson wall
沉井法　open caisson method
沉井基础　open caisson foundation, sunk caisson foundation, sunk well foundation
沉井下沉　sinking of open caisson
沉落取样器　drop sampler
沉埋式隧道　immersed tunnel
沉排　fascine mattress
沉排护岸　fascine revetment
沉砂　sand setting
沉砂池　sand and gravel trap
沉箱　box caisson, caisson

沉箱工作室	working chamber of caisson
沉箱基础	caisson foundation, pneumatic caisson foundation
沉桩	penetration of pile
衬片取样器	foil sampler
衬砌	lining
衬砌夹层防水	sandwich waterproofing of lining
衬砌隧洞	lined tunnel
衬砌止水板	water stop plate for lining
撑梁	wales
成层土	layered soil
成层岩	stratified rock
成层岩石	bedded rock
成孔灌注同步桩	synchronous pile
成孔直径	drilling diameter
成土作用	pedogenesis
成岩作用	diagenesis
成组张拉	stretching in batches
承锤头	drive head
承剪能力	shear carrying capacity
承压垫板	bearing pad
承压垫层	bearing course
承压含水层	artesian aquifer
承压破坏	bearing failure
承压水	confined water, artesian water
承压水流	artesian flow
承压水头	confined water head
承载板	loading plate
承载比	bearing ratio
承载力	bearing value, load bearing capacity
承载力安全系数	safety factor for bearing capacity
承载力基本值	basic value of bearing capacity
承载力系数	bearing capacity factor

承载力值的深宽修正	depth and width corrections for bearing capacity value
承载力值的统计修正	statistical corrections for bearing capacity value
承载面积	loaded area
承载能力	bearing capacity, load-bearing capacity
承载压力	bearing pressure
城市废料	municipal waste
城市填土	municipal landfill
程序设计	programming
持力层	bearing stratum, load bearing layer, load bearing stratum, supporting course
尺寸效应	size effect, scale effect
齿墙	key wall
赤平投影	stereographic projection
赤铁矿	hematite
充气结构物	air-supported structures
充盈系数	fullness coefficient
冲程	ram stroke
冲锤	impact hammer
冲沟	coombe, gully
冲洪积扇	pluvial alluvial region
冲击波	shock wave
冲击层灌浆	alluvium grouting
冲击地层	alluvial formation
冲击荷载	impact load
冲击块	impact block
冲击能	impact energy
冲击强度试验	impact strength test
冲击强度指数	impact strength index
冲击式破碎机	impact crusher

冲击式潜孔钻机	down-hole percussive drill
冲击式一字形钻头	chisel bit
冲击系数	impact factor
冲击压实	impact compaction
冲击压碎值试验	impact crushing value test
冲击中心	impact center
冲击钻打的竖井	shot-drilled shaft
冲击钻机	percussion drill
冲击钻进	percussion drilling
冲击钻孔	percussion borehole
冲击钻探	percussion boring
冲击钻头	chopping bit
冲积	alluviation
冲积层	alluvium
冲积层地下水	alluvial ground water
冲积地	alluvion
冲积阶地	alluvial terrace
冲积平原	alluvial plain
冲积裙	alluvial apron
冲积扇	alluvial fan, talus fan
冲积土	alluvial
冲积土壤	alluvial soil
冲积物	alluvial deposit
冲剪	punching shear
冲剪破坏	punching shear failure
冲剪试验	punch shear test
冲剪应力	punching shear stress
冲孔	punching hole
冲切承载力	punch capacity
冲切面积	punching shear area
冲蚀	washout
冲蚀性	erodibility
冲刷	outwash, wash, scour
冲刷力	scouring force

冲刷深度　depth of scour
冲洗样　wash sample
冲洗钻孔　wash boring
冲洗管　wash pipe
冲抓法　percussion and grabbling method
抽泥筒　bailer
抽砂筒　sand bailer
抽水减压井　pumping depression well
抽水试验　pumping test
抽水试验　water pumping test, well pumping test
抽水试验孔　water pumping test hole
抽水下降面积　area of pumping depression
稠度　consistency, degree of consistency
稠度计　penetrometer
稠度界限　consistency limit
稠度试验　consistency test
稠度指数　consistency index, relative consistency, relative plasticity index
出浆口　grout outlet
出口流速　outlet velocity
出水量　yield of water
出逸点　release point
初波　primary wave
初步勘察　preliminary investigation
初步勘探　preliminary prospecting
初参数法　initial parameter method
初凝　initial setting
初偏心　initial eccentricity
初期固结　preliminary consolidation
初期微震　preliminary tremor
初始超载压力　initial overburden pressure
初始沉降，瞬时沉降　immediate settlement
初始固结　initial consolidation
初始固结压力　initial consolidation pressure

初始含水量 initial moisture content
初始剪切模量 initial shear modulus
初始剪切阻力 initial shear resistance
初始剪应力比 initial shear stress ratio
初始建筑条件 initial placement condition
初始可松性系数 initial looseness coefficient
初始孔隙比 initial void ratio
初始孔隙水压力 initial pore water pressure
初始黏聚力 origin cohesion
初始破坏 incipient failure
初始强度 initial strength
初始切线模量 initial tangent modulus
初始缺陷 initial imperfection
初始塑性流动 initial plastic flow
初始条件 initial condition
初始压缩 initial compression
初始液化 initial liquefaction, primary liquidation
初始应力 initial stress, primary stress
初速度加载法 loading method of initial velocity
初位移加载法 loading method of initial displacement
初硬化 pre-hardening
初震 primary earthquake
初值问题 initial value problem
储水系数 storage coefficient
触变性 thixotropy
触变泥浆 thixotropic fluids
触变岩 contact-altered rock
触探 penetration sounding
触探法 sounding method
触探锥 penetrating cone
穿透性 penetrability
传感器 sensor, tranducer

传力锚固应力　stress in prestressing tendon at transfer or anchorage
船只或排筏冲击力　impact load of ship or raft
吹填土　dredger fill
垂直度　perpendicularity
锤击沉桩　penetration of pile with hammer blow
锤击数　number of blows
锤击应力　stress due to hammer driving, stresses due to hammer driver
纯剪　pure shear
纯砂　clean sand
纯水泥浆　straight cement mortar
磁力仪　magnetometer
磁铁矿　magnetite
磁性沉降计　magnetic probe extensometer
磁性勘探　magnetic prospecting
次固结　delayed consolidation, secondary consolidation
次固结沉降　secondary consolidation settlement
次固结速率　rate of secondary consolidation
次固结系数　coefficient of secondary consolidation
次棱角土粒　subangular particles
次生黄土　redeposited loess
次生结构面　secondary structure
次生矿物　secondary mineral
次生黏土　secondary clay
次压缩　delayed compression
次要地质细节　minor geologic details
次应力　secondary stress
次主应力　secondary principal stress
粗糙度　asperity, roughness
粗糙度测定计　profilometer

粗糙系数　roughness factor
粗粉土　coarse silt
粗骨料　coarse aggregate
粗颗粒的　macroscopic granular
粗颗粒粒度　coarsely granular particle size
粗砾　coarse gravel
粗砾质石　cobbly soil
粗粒砂　torpedo sand
粗粒砂岩　kern stone
粗粒土　coarse-grained soil
粗粒组合量　coarse fraction content
粗砂　coarse sand, grit
粗砂砾质土　gritty soil
粗砂岩　grit stone
粗筛　coarse screen
粗石蜡　paraffin wax
粗碎石　crushed broken stone
粗纹泥　megavarve
粗岩屑　coarse waste
促凝剂　coagulator
催化剂　accelerator
脆性黏土　brittle clay
脆性破坏　brittle failure
脆性破裂　brittle fracture
脆性岩石　brittle rock
脆性指数　brittleness index
搓揉　kneading
搓揉压实　kneading compaction

D

搭叠板桩　shiplap sheet piling
搭接桩墙　secant pile wall

达西定律 Darcy's law
打板桩 driving of sheet piling
打夯 ramming
打夯机 ramming machine
打夯深度 ramming depth
打夯速度 tamping speed
打梅花桩 staggered piling
打入包壳桩 driven shell pile
打入灌注桩 driven in-situ pile
打入式开缝管取土器 soil-sampling device
打入式托换桩 driven underpinning piles
打入试验桩 driven test pile
打入预制混凝土桩 driven precast concrete pile
打入桩 driven pile, displacement pite
打入桩摩阻力 skin friction of driven pile
打竖井 shaft-sinking [caisson works]
打桩 pile driving, stake out, piling
打桩船 pile driver barge(boat)
打桩锤 monkey
打桩动阻力 dynamic pile driving resistance
打桩方法 pile-driving method, piling method
打桩分析器 pile driving analyser
打桩工程 pile work
打桩公式 pile driven formula, pile driving formula
打桩机 pile driver, piling rig, ram engine, ram machine, sheeting driver
打桩机导向柱 pile driver lead
打桩记录 driving record, pile driving record
打桩架 pile driving frame
打桩落锤 drop pile hammer
打桩能量 driving energy
打桩平台 pile driving platform
打桩起重机 pile driver crane

打桩设备	driving rig, pile driving equipment
打桩深度	driving depth
打桩时的张拉力	tension stress during driving
打桩试验	driving test, pile driving test
打桩台	driving pile abutment
打桩套管	former
打桩效率	pile driving efficiency
打桩应力	pile driving stresses, driving stress
打桩噪声	piling noise
打桩阻力	driving resistance, pile driving resistance
打桩钻机工作面	working level of piling rig
大坝基础灌浆	dam foundation grouting
大比例尺土壤图	large-scale soil map
大比例模型	large-scale model
大地构造学	geotechnical geology, geotechnics
大地纬度	topographic latitude
大放脚埋深	depth of footing
大规模野外试验，大型现场试验	large scale field test
大孔结构	macroporous structure, open grain structure
大孔径钻机	large-diameter drilling machine
大孔隙	macropore
大孔隙比	macrovoid ratio
大孔隙系数	macropores coefficient
大孔性土	macroporous soil
大块卵石填沙基础	boulder foundation
大块石基层	telford base
大块式基础	massive foundation
大理石	griotte, marble
大陆冰川	continental glacier
大陆沉积	continental sedimentation
大陆架	continental shelf

大陆坡　continental slope
大卵石　cobstone
大面积屈服　large region yield
大挠度理论　large deflection theory
大气沉积物　aerial sediment
大气监测仪　air-monitoring device
大气圈　atmosphere
大气水　atmospheric water, meteorological water
大气压力　atmospheric pressure
大石块基层道路　telford road
大石块基层碎石路面　telford pavement
大体积刚性结构　massive and rigid structure
大头圆柱桩　belled-out cylindric(al) pile
大头桩　express pile
大型土工试验　large scale soil test
大应变　large strain
大应变幅　large strain amplitude
大直径扩孔灌注桩　under-reamed bored pile of large diameter
大直径双层取芯管　large-diameter double tube
大直径桩　cylinder pile
大直径钻机　large-diameter drill rig
大直径钻孔　large bore
大主平面　major principal plane
大主应变　major principal strain
大主应力　major principal stress
代用材料　alternate material
带负荷加固法　strengthening of structure under loading
带翼桩　pile with wings
带状黏土　banded clay, bandy clay, book clay, ribbon clay
带状预制排水板　band shaped prefabricated

drain

袋装砂井 fabric-enclosed (sand) drain, packed drain, sand wick, wick drain
单侧坡面墙 talus wall
单侧水压力 one-side water pressure
单纯形法 simplex method
单独基础 individual footing
单根碎石桩 individual stone column
单管桩 monotube pile
单价离子 monovalent ion
单剪 simple shear
单剪试验 simple shear test
单剪仪 simple shear apparatus
单孔沉井 single hole open caisson
单孔法 single hole method
单粒结构 single-grained structure
单锚式板桩 single anchored sheet pile
单面剪切试验 single-shear test
单面排水 one-way drainage
单排灌浆帷幕 single line grout curtain
单排孔沉井 single row hole open caisson
单位变形 unit deformation
单位表面摩擦力 unit skin friction
单位侧阻 unit shaft resistance
单位端阻力 point resistance pressure
单位法向矢量 unit normal vector
单位夯击能量 unit tamping energy
单位荷载 unit load
单位面积 unit area
单位面积荷载 unit area loading
单位体积膨胀 unit volume expansion
单位体积装药量 specific charge
单位体力 specific body force
单位吸水量 specific water absorption

单位应变　unit stain
单位应力　unit stress
单向固结　unidimensional consolidation
单液法　one-shot method
单液灌浆　single-shot grouting
单一粒径的　like-grained
单元槽段　unit trench section
单轴抗拉强度　uniaxial tensile stress
单轴抗拉强度试验　uniaxial tension strength test
单轴拉力　uniaxial stress
单轴向压缩　uniaxial compression
单轴压力　uniaxial compression force
单轴压缩试验　uniaxial compression test
单桩　single pile
单桩沉降量　settlement of single pile
单桩承载力　bearing capacity of single pile
单桩横向极限承载力　ultimate lateral resistance of single pile
单桩横向受力分析的常系数法
　　constant coefficient method of lateral loading analysis of single pile
单桩极限承载力　ultimate carrying capacity of single pile
单桩竖向静载荷试验　static load test of pile
单桩竖向抗拔极限承载力
　　vertical ultimate uplift resistance of single pile
单桩竖向抗压极限承载力
　　vertical allowable load capacity of single pile, vertical ultimate carrying capacity of single pile
弹簧锥贯入度计　spring cone penetrometer
弹塑性　plasto-elasticity
弹塑性变形　plastic-elastic deformation
弹塑性范围　elastoplastic range

弹塑性分析	elastic-plastic analysis
弹塑性矩阵	elastic-plastic matrix
弹塑性理论	elasto-plasticity theory
弹塑性模型	elastic-plastic model
弹塑性耦合	elastoplastic coupling
弹塑性体	elastic-plastic solid, plastic-elastic body
弹塑性状	elastic-plastic behavior
弹性半空间地基模型	elastic half-space foundation model
弹性半空间理论	elastic half-space theory
弹性半无限地基	elastic semi-infinite foundation
弹性半无限体	elastic semi-infinite body
弹性边墙	elastic side wall
弹性变形	elastic deformation, recoverable deformation, recovery of elasticity
弹性表面波	surface elastic wave
弹性薄壳理论	theory of elastic thin shell
弹性常数	elastic constant
弹性沉降	elastic settlement
弹性地基	elastic foundation
弹性地基梁	beam on elastic foundation
弹性地基梁比拟法	beam on elastic foundation analogy, elastic foundation beam analogy method
弹性地基梁法	elastic foundation beam method
弹性垫层	elastic fill
弹性非均匀压缩系数	coefficient of elastic non-uniform compression
弹性公式	elastic formula
弹性灌浆料	elastomer
弹性后效	delayed elasticity, elastic after effect
弹性基床反力	elastic subgrade reaction

弹性畸变　elastic distortion
弹性极限　elastic limit
弹性剪切系数　coefficient of elastic shear
弹性介质　elastic medium
弹性均匀压缩系数　coefficient of elastic uniform compression
弹性理论　elastic theory
弹性力　elastic force
弹性梁　elastic beam
弹性隆起　elastic heave
弹性模量　elastic modulus, elasticity modulus, modulus of elasticity, Young's modulus
弹性模量比　ratio of elastic modulus
弹性平衡　elastic equilibrium
弹性平衡状态　elastic state of equilibrium, state of elastic equilibrium
弹性区　elastic zone
弹性体　elastic solid
弹性压缩　elastic compression
弹性应变　elastic strain, rebound strain
弹性应变能　elastic strain energy
弹性支座　elastic support
淡水　fresh water
当量半径　equivalent radius
当量筛孔尺寸　equivalent opening size(EOS)
当量相对密度　equivalent relative density
挡水坝　retaining dam
挡土板　lagging, retaining slab
挡土构筑物　retaining structure
挡土墙　retaining wall, abamurus
挡土墙台阶基础　stepped footing
挡土墙稳定性　stability of retaining wall
刀脚　cutting edge

刀口	knife edge
导板	guide runner
导洞	guide adit, heading
导沟	guide trench
导管	conductor pipe, guide tube
导管法	tremie method
导架	lead
导坑	approach pit
导坑法	pilot tunnel method
导流洞	diversion tunnel
导墙	guide wall
导热率	conductivity
导水系数	coefficient of transmissibility, transmissivity
导桩	guide pile, leading pile, pile guide, pilot pile
倒拱基础	inverted arch foundation
倒梁法	inverted beam method
捣锤冲击机	ram impact machine
捣实黏土	pise, puddle clay
到达历时	arrival time
道碴	ballast
道路基层	base course
道路路面	road pavement
稻草板	rice-straw board
稻草灰	rice-straw ash
德佛特连续取样器	Delft continuous sampler
德鲁克-普拉格准则	Drucker-Prager criterion
等沉面	plane of equal settlement
等沉曲线	curve of equal settlement
等地温线	geoisotherms, isogeotherm
等冻结线	freezing isoline
等高线图	contour map
等厚线	isopachyte

等强度 uniform strength
等强度曲线 iso-intensity curve
等强度线图 strength contour graph
等倾线 isoclinic
等色线照片 isochromatic photograph
等深线 bathymetric line
等渗压线 isosmotic pressure line
等湿度线 isohume
等时孔压线 isochrone
等势线 equipotential line
等水深浅 hydroisohypse
等速贯入试验 constant rate of penetration test
等速加荷固结试验 consolidation test under constant loading rate
等速上拔试验 constant rate of uplift test
等梯度固结试验 constant gradient test
等体积波 equivoluminal wave
等温线 isotherm
等效比热 equivalent specific heat
等效材料 equivalent material
等效固结压力 equivalent consolidation pressure
等效荷载原理 principal of equivalent loads
等效基础模拟 equivalent footing analogy, footing analogy
等效孔径 equivalent opening size(EOS)
等效粒径 equivalent diameter, equivalent grain size
等效流体 equivalent fluid
等效内摩擦角 equivalent angle of internal friction
等效应力 equivalent stress
等效阻尼比 equivalent damping ratio
等压线 equipressure lines, isobar, isostatic

curve
等应变速率固结试验　consolidation test under constant rate of strain
等值层法　equivalent soil layer method
等值梁法　equivalent beam method
等值线　contour line, isoline
等值线法　isoline method
等重量代替法　equivalent weight replacement method
低承台基桩　pile foundation with low cap
低地下水位　phreatic low
低强度水泥　low strength cement
低填土　shallow fill
低桩承台　low pile cap
堤　dike, dyke, levee
堤岸加固　bank strengthening, bank stabilization
堤防背水面　back of levee
堤防维修　dyke maintenance
堤防隐患探测　dyke defect detecting
堤防钻探　dyke boring
堤基　base of levee
堤面护坡　levee revetment
滴定管　burette
底板　baseplate, seat earth
底板式基础　bed-plate foundation
底部掏槽　toe cut
底压力　bottom pressure
底座　bedplate
抵抗变形　resistance to deformation
抵抗力　resisting power
地表　ground surface
地表水　open water, surface water
地表植被　ground vegetation

地槽	geosyncline
地层	stratum, formation
地层的	stratigraphic
地层横剖面	stratigraphic cross section
地层控制	strata control
地层锚杆	ground anchor
地层倾斜角	angle of bedding
地层渗透率	in-place permeability
地层时序表	chronological time scale
地层压力	formation pressure, ground pressure
地动	ground movement
地冻现象	frozen-ground phenomenon
地沟	ground sluice, trench
地鼓	swelling of floor
地基	foundations, ground base, subbase
地基沉降	foundation settlement
地基沉降量	settlement of subsoil
地基承载力	bearing capacity of foundation, ground bearing pressure
地基处理	ground treatment, foundation treatment
地基冻结法	ground freezing
地基反力动力系数	coefficient of dynamic(al) subgrade reaction
地基反力系数	modulus of subgrade reaction, coefficient of subgrade reaction
地基工程	ground engineering, foundation work
地基固结	ground consolidation
地基灌浆	foundation grouting
地基回弹	rebound of foundation
地基回弹测试	foundation resilience test
地基极限承载力	ultimate bearing capacity of

	subsoil
地基加固	consolidation of soil, foundation stabilization, ground improvement, ground stabilization
地基剪切破坏模式	modes of shear failure for subsoil
地基开挖	foundation excavation
地基勘察	foundation investigation
地基勘探	foundation exploration, subsurface exploration
地基抗弯刚度系数	rocking stiffness factor
地基抗压刚度系数	vertical stiffness factor
地基梁	ground beam
地基临界荷载	critical load of subsoil
地基模量	foundation modulus
地基排水	drainage of foundation
地基容许承载力	allowable bearing capacity of foundation soil
地基设计	foundation design
地基渗漏	foundation leakage
地基条件	ground condition
地基土	foundation soil
地基土层	foundation strata
地基土极限荷载	ultimate load of subsoil
地基稳定性	stability of subsoil
地基压力扩散角	spread angle of pressure in subsoil
地基压缩	ground compaction
地基压缩层厚度	thickness of compressed layer of foundation
地基应力	stress in soil mass
地基允许变形值	allowable deformation of subsoil
地脚螺栓	foundation bolt

地壳　earth crust
地壳运动　earth movement
地沥青　asphalt
地裂　ground fracturing
地貌　landform
地貌　topography
地貌单元　landform unit
地貌学　geomorphology
地面标高　ground level
地面沉降　ground settlement, land subsidence
地面沉陷破坏　afterbreak
地面护坡　sheeting planks of dam
地面加速度　ground acceleration
地面隆起　land upheaval
地面排水　land drainage
地面坡度　inclination of ground
地面倾斜因素　ground slope factor
地面塌陷　surface subsidence
地面下沉　ground subsidence
地面下陷　ground loss
地面运动　ground motion
地球外的土　extraterrestrial soil
地球物理测井　geophysical log
地球物理勘探　geophysical exploration
地球物理学　geophysics
地区性土　regional soil
地势　terrain
地文学　physiography
地下　underground
地下爆破　underground blasting
地下硐室　underground opening
地下工程　underground engineering, underground work
地下灌浆工程　underground grouting works

地下河	estavel
地下径流	runoff in depth, subsurface runoff
地下开挖	underground excavation
地下勘探	underground exploration
地下连续墙	continuous concrete (slurry) wall, underground diaphragm wall
地下排水工程	mole drainage
地下排水沟	mole drain
地下潜蚀	underground erosion
地下侵蚀破坏	failure by sub-surface erosion
地下渗流	underground leakage, subsurface seepage flow, underseepage
地下室	underground chamber
地下室防水	basement water proofing
地下室墙	basement wall
地下水	phreatic water, underground water, groundwater
地下水补给量	groundwater increment, groundwater recharge
地下水不稳定流	groundwater unsteady flow
地下水产出条件	occurrence condition of groundwater
地下水等水位线图	contour map of groundwater
地下水等位线	groundwater isopiestic line
地下水动力学	groundwater dynamics
地下水动态	groundwater regime
地下水分水岭	groundwater divide
地下水分水线	groundwater ridge
地下水跟踪试验	groundwater tracer test
地下水监测	groundwater monitoring
地下水控制	control of underground water, groundwater control
地下水矿化度	mineralized degree of groundwater

地下水流的电模拟　electrical analogue for ground water flow
地下水流量　phreatic discharge
地下水流域　drainage basin of groundwater
地下水排泄　groundwater discharge
地下水瀑布　groundwater cascade
地下水人工补给　artificial recharge of ground water
地下水上部含水层　phreatic high
地下水上升　phreatic rise
地下水实际流速　field groundwater velocity
地下水位　groundwater elevation, groundwater level(table), underground water level, water table
地下水位变动带　fluctuation belt of water table
地下水位漏斗线　cone of water-table depression
地下水位下降　decline of underground water level, phreatic decline
地下水位线　phreatic line
地下水稳定流　groundwater steady flow, ground water pollution
地下水污染　pollution of ground water
地下水硬度　groundwater hardness
地下水蒸发　groundwater evaporation
地下水贮存量　groundwater storage
地下水总矿化度　total mineralization of groundwater
地下隧道　underground tunnel
地下炸药库　underground store for explosives
地形　physical landform
地形测量　topographic survey
地形单元　landform element
地形点　topographic point

地形勘探	topographic reconnaissance
地形图	ground map, topographic map
地形线	terrain line
地形状况	topographic condition
地压力	geostatic pressure, geostress
地应力	ground stress
地应力观测	observation of ground stress
地震	earthquake
地震波	earthquake wave
地震波传播路径	path of seismic waves
地震测井	bore hole seismic logging, bore-hole seismic logging
地震查勘	seismic survey
地震动土压力	earthquake dynamic earth pressure
地震断层	earthquake fault
地震反应谱	earthquake response spectrum
地震工程学	earthquake engineering
地震勘探	seismic exploration, seismic prospecting
地震力	seismic force
地震烈度	earthquake intensity
地震烈度表	earthquake intensity scale
地震前兆	preliminary symptom
地震效应	earthquake effects
地震影响系数	coefficient of seismic effect
地震震级	earthquake magnitude
地震卓越周期	seismic predominant period
地质报告	geological report
地质不连续面	geological discontinuity
地质成因	geological origin
地质锤	geologists' pick, prospecting hammer
地质点	geological observation point
地质调查	geological survey

地质构造	geologic structure, geological structure
地质环境	geologic environment
地质环境要素	geologic environment element
地质记录	geological log
地质勘察	geologic investigation
地质勘探	geological exploration, geological prospecting
地质力学	geomechanics
地质力学模型试验	geomechanical model test
地质历史	geological history
地质年代	geological age, geological period
地质年代学	geological chronology
地质剖面图	geological profile, geological section
地质缺陷	geological defect
地质水文学	geohydrology
地质特征	geologic features
地质条件	geological condition
地质图	geological map
地质学	geology
地质学家	geologist, rock hound
地质演变	geological succession
地质柱状图	geological column
第二类失稳	instability of second kind
第三纪	Tertiary
第四纪	alluvium period, quaternary
第四纪冰川	quaternary glacier
第四纪沉积物	quaternary deposit
第四纪地质图	quaternary geological map
第四纪土质学	quaternary geology
第四季冲积层	Quaternary alluvium
第一类失稳	instability of first kind
点荷法	the method of point load

点荷载试验　point loading test
点式记录仪　dot recording system
电测井　electric log
电测孔隙水压力仪　electrical piezometer
电测剖面法　electric profiling
电测深　electrical sounding
电测水位计　electric fluviograph
电磁打桩机　magnetoelectric pile driver
电磁地下测探　electromagnetic subsurface probing [ESP]
电磁勘探　electromagnetic prospecting
电磁桩锤　magnetoelectric pile hammer
电动排水　electrical drainage
电法测井　resistivity log
电法勘探　electric prospecting, electrical prospecting
电化学加固　electrochemical stabilization
电化学效应　electrochemical effect
电加固　electric stabilization
电离层　ionization layer, ionized layer
电离势　ionization potential
电模拟法　electrical analog method
电渗法　electro-osmosis
电渗加固　consolidation by electroosmosis, electro-osmotic stabilization
电渗排水　drainage by electro-osmosis, electrical-drainage
电位测井　potential logging
电阻法　resistivity exploration
电阻率　resistivity
电阻率法　resistivity method
垫板　rails tie plate
垫层　bedding cushion, mattress, cushion
垫层材料　bedding(course) material

垫层混凝土　bedding concrete
垫层砂浆　bedding mortar
垫层应力系数　cushion stress factor
垫底　underlaying bed
垫木　kicking piece
垫式基础　pad foundation
淀积土壤　illuvial soil
调凝水泥　regulated set cement
叠加法　method of superposition
叠加原理　principal of superposition
叠片体　domain
叠置桩　overlapping pile
碟形沉降　bowl-shape settlement
丁坝　groin
钉桩　dowel pile
顶板　crown plate
顶板控制　roof control
顶部掏槽　top cut
顶管法　conduit jacking
顶管法　pipe jacking method
顶梁　jacking beam
顶升法　jacking method
顶推法施工　incremental launching method
顶样器　extruder
定比例　scaling
定灌喷浆　directional jet grouting
定线桩　setting out rod, setting stake
定向　orientation
定向剪切盒　directional shear cell
定向钻进　directional drilling
动承载力　dynamic bearing capacity
动单剪试验　dynamic simple shear test
动荷载　dynamic load
动基床反力　dynamic subgrade reaction

动抗剪强度　dynamic shear strength
动孔隙压力　dynamic pore pressure
动力测桩　dynamic pile test
动力触探　dynamic sounding
动力触探锤击　blow of dynamic sounding
动力触探试验　dynamic penetration test
动力打桩公式　dynamic-pile driving formula
动力弹性模量　dynamic elastic modulus
动力反应　dynamic response
动力反应法　dynamic response approach
动力放大因素　dynamic magnification factor
动力固结仪　dynamic oedometer
动力夯　power rammer
动力互换　dynamic reciprocity
动力黏滞系数　coefficient of dynamic viscosity
动力黏滞性　dynamic viscosity
动力黏滞性　kinematic viscosity
动力探头阻力　dynamic point resistance
动力稳定，动力平衡　kinetic stability
动力有效应力法　dynamic effective stress method
动力圆锥触探仪　dynamic cone penetrometer
动力置换法　dynamic replacement
动摩擦　kinetic friction
动三轴试验　dynamic triaxial test
动水位　dynamic level
动态分析　dynamic analysis
动态岩石强度　dynamic strength of rock
动态移动　dynamic displacement
动应力　dynamic stress
动阻力　dynamic resistance
冻附力　adfreezing force
冻附强度　adfreezing strength
冻结层间水　interpermafrost water

冻结带 freezing zone
冻结法 freezing method, freezing process
冻结力 freezing force
冻结前缘 frost front
冻结强度 frost strength
冻结区 frost zone
冻结取样器 freezing sampler
冻结深度 depth of frost penetration, frost depth, frost penetration
冻结指数 freezing index
冻结作用 frost action, frost effect
冻融 frost boil
冻融固结 thaw consolidation
冻融试验 freeze-thaw test
冻土 frozen soil, tjaele, tundra
冻土抗剪强度 shear strength of frozen soil
冻土融化压缩试验 thaw compression test of frozen soil
冻胀 frost heave
冻胀力 frost heaving pressure, frozen-heave force, frost-heaving pressure
冻胀量 frost-heave capacity
冻胀土 frost-heaving soil
冻胀压力 frost pressure
冻胀仪 frozen heave test apparatus
峒室稳定性评价 evaluation of opening stability
洞室围岩变形监测 monitoring of surrounding rock deformation of tunnel
洞穴 cavern
洞穴沉积 cave deposit
陡岸 overhanging bank
陡坡 abrupt slope, heavy grade, rapid ascent, steep slope
陡坡墙 scarp wall

陡倾	steep dip
陡崖	escarpment
豆石	pisolite
独立基础	individual footing, isolated foundation
度-日(冻土用单位)	degree-days
端部约束效应	end restraint effect
端承摩擦桩	combined end-bearing and friction pile
端承柱	end-bearing pile, end bearing pile, point bearing pile
端垫误差	bedding error
端阻力	end bearing
短期荷载试验	short term loading test
短期荷载应力	stress for temporary loading
短期瞬时荷载	short-term transient load
短圆柱壳	short cylindrical shell
短支柱	short post
短桩处理	treatment with short pile
断层	broken course, fault, dislocation
断层带	fault zone
断层角砾石	crust breccia
断层面	fault plane
断层泥	fault gouge, gouge
断层石块	fault block
断裂	rupture, fracture
断裂力学	fracture mechanics
断裂面	fracture plane, plane of break, plane of rupture
断裂试验	breaking down test
断裂收缩率	reduction of area at fracture
断裂应变	breaking strain
断裂应力	breaking stress
断桩高程	cut-off elevation

锻锤基础　forcing hammer foundation
堆积阶地　accumulation terrace, construction terraces
堆积密度　apparent density
堆积区　accumulation area
堆积台地　accumulational platform
堆石坝　rockfill dam
堆石防波堤　all rock breakwater
堆石护坡　riprap stone revetment, rock facing
堆填法　stack filling
堆载加压纠偏　tilting correction by ballast loading
堆载预压　preloading
对称面　plane of symmetry
对数比例尺　log scale
对数减量　logarithmic decrement
对数螺线　logarithmic spiral
墩式基础　pier foundation
盾构　shield
盾构法　shield driving method, shield method
盾构法隧道　shield tunnel
盾构法挖掘机　shield tunneling machine
盾构纠偏　shield steering adjustment
盾构掘进　shield driven, shield excavation
盾构灵敏度　sensibility of shield, shield sensitivity
盾构密封间隙　sealing clearance at the tail of shield
盾构千斤顶　shield jack
盾构施工法　shield construction
盾构式掘进隧道　shield-driven tunnel
盾构同步压浆　simultaneously grouting
盾壳　shield shell
盾尾　shield tail, tail

盾尾密封装置　sealing device at the tail of shield
多层半无限体　multilayer semi-infinite solid
多层薄膜　multi-membrane
多层加强　multi-reinforcement
多层锚定系统　multi-anchorage system
多次反射　multiple reflection
多点应变计　multiple point extensometer
多斗挖槽机　ladder trencher
多格基础　egg box foundation
多级井点系统　multi-stage well point system
多级排水　multi-stage drainage
多级三轴压缩试验　multi-stage triaxial compression test
多级套筒式沉井　Gow caisson
多角骨料　sharp aggregate
多节桩　multi-section pile
多孔灌浆记录仪　multiple injection recorder
多孔集料　porous aggregate
多孔结构　porous structure
多孔介质　porous medium
多孔排水管　perforated drain pipe
多孔套管　perforated casing
多孔土　porous soil
多孔性地基　porous foundation
多孔岩石　spongy rock
多能三轴仪　multi-purpose triaxial test apparatus
多年冻土　perennial frozen soil
多排孔沉井　multi-well open caisson
多排帷幕　multiple line curtain
多缺陷桩　multi-defective pile
多头测压管　multiple piezometer
多维固结　multi-dimensional consolidation
多相的　multiphase

多向剪切　multi-directional shear
多用钻孔　polydrill
多重应力集中　multiple stress concentration
多轴拉伸试验　multiaxial tensile test

E

二次爆破　secondary blasting
二次地压　steady secondary ground pressure
二次灌浆　back grouting
二次浇筑　secondary cast
二次膨胀　secondary expansion
二次压注　secondary grouting
二次注浆　regrouting
二次注浆法　twice grouting method
二迭纪　Permian period
二灰土桩复合地基　composite ground with lime-fly ash columns
二维固结　two dimensional consolidation

F

发射触探　projectile penetration
筏形基础　buoyant foundation, mat foundation, raft foundation
法兰基灌注桩　Franki pile
法向刚度　normal stiffness
法向力　normal force
法向压力　normal pressure
法向压应力　normal compressive stress
法向应变　normal strain
法向应力　normal stress

翻浆　mudding
翻浆冒泥　mud pumping
反铲挖土机　backhoe
反冲力　recoil
反复弯曲疲劳　repeated bending fatigue
反复胀缩强度　shear strength of repeated swelling shrinkage
反复直剪强度试验　repeated direct shear test
反馈控制　feedback control
反滤层　filter, inverted filter, protective filter
反滤铺盖　filter blanket
反坡　dip slope, retrogressive slope
反射法勘探　reflection survey
反渗压力　backosmotic pressure
反向铲　back digger
反向工作法　reverse work
反向弯矩　reverse bend
反向弯矩试验　reverse bend test
反向压力　reverse compression
反絮凝剂　deflocculating agent
反絮凝作用　anti-flocculation, deflocculation
反压护道　loading berm
反压力　anti-vacuum, back pressure
反演分析　inverse analysis, back analysis
反应曲线　bearing graph
反作用力　reaction force
反作用扭矩　torque reaction
泛滥平原沉积　flood plain deposit
范德华力　Van der Waals force
方程余量　residue of equation
方解白云石　calcite dolomite
方解大理石　calcite marble
方解石　calcareous spar, calcite, calcspar
方形独立基础　square spread footing

方形基础	square footing
方钻杆	kelly bar
防波堤堤头部	pierhead
防潮	damp proofing
防潮层	damp proof course, moisture barrier
防冲墙	anti-scour wall
防冻深度	frost-proof depth
防辐射墙	radiation wall
防腐处理	preservative treatment
防腐涂层	anti-corrosion protection coat
防漏的	leakproof
防燃土工布	flame retardant fibers
防渗	seepage prevention
防渗膜	anti-seep diaphragm
防渗铺盖	impervious blanket
防渗体	impervious barrier
防渗帷幕	seepage proof curtain, impervious curtain
防水的	watertight
防水混凝土	waterproof concrete
防水接缝	watertight joint
防水心墙	water tight core
防水性	watertightness
防水药包	waterproof charge
防塌工程	avalanche prevention works
防雨篷	rain canopy
防震缝	earthquake proof joint
防震基础	vibration-proof foundation
纺成土工织物	woven fabrics(geotextile)
放大系数	magnification factor
放大因素	amplification factor
飞石	scattering stone
飞行图	aeronautical chart
非饱和孔隙度	unfilled porosity

非饱和流	unsaturated flow
非饱和试验	unsaturation test
非饱和土	unsaturated soil
非保守力	no conservative force
非层状土	unstratified soil
非承压地下水	unconfined groundwater
非承压水体	unconfined water body
非弹性变形	inelastic deformation
非弹性碰撞	inelastic collision
非弹性屈曲	inelastic buckling
非地震区	non-seismic region
非构造地裂缝	non-tectonic ground crack
非构造断层	non-tectonic fault
非构造节理	non-tectonic joint
非关联流动法则	non-associative flow rule
非级配混合料	nongraded mix
非晶质	amorphous
非晶质黏土矿物	noncrystalline clay mineral
非均匀流	nonuniform flow
非均质地层	heterogeneous stratum
非均质土	heterogeneous soil, inhomogeneous soil, nonhomogeneous soil
非均质性	heterogeneity, inhomogeneity
非颗粒性灌浆	non-particulate grout
非黏土矿物	non-clay mineral
非黏性的	non-cohesive
非破坏性测试	non-destructive testing(NDT)
非破损试验	non-destructive test
非湿化性黏土	non-slaking clay
非湿陷性土	non-collapsing soil
非塑性的	non plastic
非完整井	partially penetrated well
非稳定流	non-stationary flow
非稳定流	unsteady flow

非稳定渗流　unsteady seepage
非线性　nonlinearity
非线性边界条件　nonlinear boundary condition
非线性大挠度稳定理论　nonlinear large deflection theory of stability
非线性弹性理论　nonlinear elasticity theory
非线性弹性模型　nonlinear elastic model
非线性弹性性状　nonlinear elastic behavior
非线性分析　nonlinear analysis
非线性理论　nonlinear theory
非线性模型　nonlinear model
非线性黏弹性模型　nonlinear viscoelastic model
非线性相关　nonlinear correlation
非线性有限元分析　nonlinear finite element analysis
非线性约束　nonlinear constraints
非圆弧分析法　non-circular analysis
非周期运动　aperiodic motion
非周期阻尼　aperiodic damping
非自流地下水　non-artesian ground water
非自流井　non-artesian well
非自重湿陷性黄土　self weight non-collapse loess
肥黏土　fat clay, gumbo
废河道　abandoned channel
废料处理　waste disposal
废料处理设备　waste disposal facilities
废料覆盖层　waste mantle
废料污染　waste containment
废料岩土力学　waste geotechnics
废泥浆　mud disposal
沸石　zeolite
费伦纽斯解法(地基承载力)　Fellenius solution

费伦纽斯条分法	Fellenius method of slices
分布荷载	distributed load
分层夯实	tamping in layers
分层黏土	varved clay
分层铺筑	layer construction
分层取样	stratified random sampling
分层填筑	placement in layers
分层现象	stratification
分层压实	compaction in layers
分层总和法	layerwise summation method
分叉理论	bifurcation theory
分段挡土墙	sectional retaining wall
分段固井技术	stage cementing technology
分段栓塞水压试验	packer test
分段张拉	tensioning by section
分割粒径	cut diameter
分级加载	loading in increment
分级湿陷量	collapsibility grading index, grading collapse settlement, wet-subsidence for classification
分级卸荷	decrementation
分解作用	decomposition
分界粒度颗粒	near-mesh grain
分界面	interface
分类	classification
分类试验	classification test
分离层	unconnected course
分期灌浆	stage grouting
分散剂	dispersant(dispersing agent)
分散结构	dispersed structure
分散系数	dispersion coefficient
分散性	dispersity
分散性黏土	dispersive clay(soil)

分散作用	disaggregation, dispersion
分土器	sample divider
分项系数	partial factor
分项系数法	partial factor method
分选系数	coefficient of sorting
分样器	sample splitter
分子键	molecular bond
分子键力	intermolecular bonding force
分子力	molecular force
分子链形态	shape of molecular chain
分子黏聚力	molecular cohesion
分子引力	molecular attraction
分组指数	group index
玢岩	porphyrite
粉煤灰	flyash
粉煤灰加固土	flyash stabilized soil
粉砂	silty sand
粉砂岩	aleurolite (siltite)
粉土	silt
粉质黏土	silty clay
风沉积	wind laid deposit
风成砂	blown sand
风成砂岩	anemoarenyte
风成岩	aeolianite
风动锤	air ram
风动发动机	air motor
风动凿岩机	pneumatic rock drill
风动钻机	air motor drill
风洞锤	air driven hammer
风干土	air-dried soil
风镐	rock picker
风化变质	katamorphism
风化层	regolith, weathered layer, mantle rock

风化程度　degree of weathering
风化带　weathering zone, weathered zone
风化花岗石　decomposed granite, weathered granite
风化壳　weathered crust, weathering crust
风化料　weathered rock-soil
风化土　saprolite
风化系数　coefficient of weathering
风化岩　weathered rock
风化岩层　decayed rock
风化岩石　decomposed rock, weathered rock
风化作用　weathering
风积土　aeolian deposit(s) (aeolian soil)
风积物　aeolian sediment, eolian deposit
风砂　wind drift sand
风砂磨蚀　abrasion of blown sand
风蚀　aeolian erosion, deflation, eolation, wind abrasion
风钻　pneumatic drill
封闭空气　entrapped air
封闭系统　closed system
封孔　hole sealing
封锚　sealing anchorage at beam end
封填土　sealed earth
峰值　peak value
峰值抗剪强度　peak shear strength
峰值孔压比　peak pore pressure ratio
峰值强度　peak strength
峰值应变　peak strain
峰值应力应变　strain corresponding to the maximum stress
蜂窝结构　honeycomb structure
缝隙　joint opening
扶垛墙　buttressed wall

扶壁式挡土墙　counterfort retaining wall
浮吊　floating crane
浮筏基础　buoyancy raft, floating foundation
浮环式固结仪　float-ring consolidometer
浮浆层　laitance
浮力　levitating force, uplift force
浮升补偿器　heave compensator
浮石　pumice
浮式沉井稳定性　stability of float caissons
浮托力　buoyancy
浮运沉井法　method of floating open caisson
浮运沉箱　floating caisson
浮重度　buoyant unit weight
浮桩　buoyant pile
辐射井　radial wells
辅助炮眼　subsidiary shot hole
腐泥　putrid mud
腐泥土　saprolitic
腐蚀　corrasion
腐蚀率　corrosion rate
腐殖泥　gyttja
腐殖土　cumulose soil, humus soil
负孔隙(水)压力　negative pore water pressure
负摩擦力　negative mantle friction
负摩擦桩　negative friction pile
负压测定仪　pressure membrane apparatus
负压钻井　underbalanced drilling
附加荷载　superload, additional load, secondary loading
附加力矩　secondary moment
附加条款　additional clauses
附加应力　additional stress, superimposed stress
附着重力　adhesive weight

复打　redriving
复合冲积扇　compound alluvial fan
复合地基　composite foundation(ground)
复合滑动面　composite sliding(slip) surface
复活断层　renewed fault
复模量　complex modulus
复式边坡　composite slope
复式加筋　compound reinforcement
副井　subsidiary shaft
富黏土　rich clay
覆盖层　covering layer, overburden layer, overlying bed
覆盖层地质图　drift map
覆盖层灌浆　overburden grouting
覆盖层厚度　depth of overburden
覆盖范围　coverage
覆盖土　mulching soil
覆盖压力　overburden pressure

G

改善级配加固法　mechanical stabilization
钙长石　biotine
钙质结核　calcareous concretion, lime concretion
钙质软泥　calcareous ooze
盖板式涵洞　slab culvert
盖帽模型　cap model
盖帽屈服模型　capped yield model
概率不定性　probabilistic uncertainty
概率分析　probabilistic analysis
概率设计　probabilistic design
干成孔灌注桩　dry-drilled pile

干捣法　dry rodding
干度　degree of dryness
干旱区　arid region
干击法　dry tamping
干裂　season crack
干密度　dry density
干砌石护坡　stone revetment
干强度　dry strength
干筛　dry screening
干湿试验　wetting and drying test
干式钻进　dry drilling
干填　dry pack
干填砂浆　drypack
干土密度　density of dry soil
干永冻层　dry permafrost
干燥器　desiccator
干燥作用　desiccation
干振成孔灌注桩　vibratory bored pile
干重度　dry unit weight
杆钻　bank drill
感应起爆　sympathetic detonation
感应图　influence chart
感应系数　influence coefficient
感应值　influence value
刚度　stiffness
刚度比　stiffness ratio
刚度系数　coefficient of rigidity, coefficient of stiffness, stiffness factor
刚塑性模型　rigid plastic model
刚性　inflexibility
刚性挡土墙　non-yielding retaining wall
刚性导向钻杆　rigid guide-rod
刚性地基　rigid base
刚性垫层　rigid fill

刚性墩　rigid pier
刚性涵洞　rigid culvert
刚性基础　rigid footing(foundation)
刚性角　pressure distribution angle of masonry foundation
刚性连接　rigid connection
刚性模量　rigidity modulus, modulus of rigidity
刚性心墙　rigid core wall
刚玉　corundum
钢板桩　metal sheet pile, steel sheet pile
钢板桩围堰　steel sheet pile cofferdam
钢沉井　steel open caisson
钢沉箱　steel pneumatic caisson
钢管桩　steel pipe pile, steel tube pile
钢轨底座　rail bed
钢筋横断面　section area of reinforcement
钢筋混凝土板桩　reinforced concrete sheet pile
钢筋混凝土沉箱　reinforced concrete caisson
钢筋混凝土挡土墙　reinforced concrete retaining wall
钢筋混凝土地基　reinforced concrete base
钢筋混凝土拱涵　reinforced concrete arch culvert
钢筋混凝土剪力墙　reinforced concrete shear wall
钢筋混凝土开口沉箱　reinforced concrete open caisson
钢筋混凝土气压沉箱　reinforced concrete pneumatic caisson
钢筋混凝土柱　reinforced concrete column, reinforced concrete pile
钢筋笼　cage of reinforcement, steel reinforcement cage
钢筋网　mesh reinforcement

钢绳冲击钻(探) cable-tool drill(ing)
钢丝网加固喷射混凝土 wire mesh-reinforced shotcrete
钢弦式压力盒 vibrating wire cell
钢弦式应变仪 vibrating wire strain gauge
钢桩 steel pile
高承台基桩 pile foundation with high cap
高程 elevation
高程基准面 seal-level datum
高低刃脚沉井 open caisson with a tailored cutting edge
高架道路,高架桥 viaduct
高阶弹性模量 high-order elastic modulus
高进气孔隙水压力测头 high air entry piezometer tip
高跨比 ratio of rise to span
高岭石 clayite, kaolinite
高岭土 kaolin
高流动性 soft flow
高炉炉渣 blast furnace slag
高斯正态分布曲线 Gauss normal distribution curve
高塑性土 highly plastic soil
高压灌浆 high pressure grouting
高压喷射注浆法 jet grouting method
高压缩黏土 highly compressed clay
高液限土 high liquid limit soil
高原 plateau
高桩承台 high-rise pile cap
戈布尔打桩分析器 Goble pile-driving analyser
割线模量 secant modulus
割线模量比 ratio of secant modulus
格筏基础 grid-mat foundation
格构梁 lattice beam

格构柱　lattice column
格里菲斯强度准则　Griffith's strength criterion
格排　crib
格排筏式基础　grillage raft
格排基础　grillage foundation
格墙沉井　open caisson with cross walls
格式岸壁　cell quarry wall
格式岸壁[钢板桩]　cellular bulkhead
格形挡土墙　cellular retaining wall
格形围堰　cellular cofferdam
格栅　bar screen
隔板　diaphragm
隔层　interlayer
隔膜泵　diaphragm pump
隔水层　impermeable layer
隔水黏土　roof clay
隔水黏土层　clay pan
隔振　isolation, vibration isolation
隔振沟　isolation trench
隔振效果　isolation effectiveness
各向不等压固结　anisotropic consolidation
各向不等压固结不排水试验
　　consolidated anisotropically undrained test
各向不等压固结排水试验
　　consolidated anisotropically drained test
各向不等压硬化　anisotropic hardening
各向等压固结　isotropic consolidation
各向等压固结不排水试验
　　consolidated isotropically undrained test
各向等应力　isotropic stress
各向同性　isotropy
各向同性半无限体　isotropic semi-infinite solid
各向同性材料　isotropic material
各向同性沉积物　isotropic deposit

各向同性土 isotropic soil
各向同性硬化 isotropic hardening
各向异性 anisotropy
各向异性土 anisotropic soil
给水度 specific yield
给水隧道 water supply tunnel
跟进套管 advanced casing
更新世 Pleistocene epoch
工程 project
工程地貌学 engineering geomorphology
工程地址勘查 site investigation
工程地质测绘 engineering geological mapping
工程地质测绘勘查报告 report of engineering geologic investigation
工程地质单元 engineering geologic unit
工程地质分区 engineering geologic zoning
工程地质勘察报告 engineering geologic investigation report
工程地质勘探 engineering geologic investigation (exploration), engineering geological prospecting
工程地质类比法 engineering geological analogy
工程地质评价 engineering geologic evaluation, engineering geological evaluation
工程地质剖面图 engineering geologic profile
工程地质师 engineering geologist
工程地质条件 engineering geological condition
工程地质图 engineering geologic map
工程地质学 engineering geology
工程地质柱状图 engineering geologic columnar profile
工程地质钻探 engineering geologic drilling
工程地质作用 engineering geologic process

工程放线　job layout
工程实录　case history
工程岩土学　rock and soil engineering
工地　job site, worksite
工地试验室　on-job laboratory, site laboratory
工后监测　post-construction monitoring
工棚　shelter
工期缩短　reduction of erection time
工业废料　industrial waste
工业污染物　industrial contaminant
工作平台　working platform
工作桩　working pile
弓形拱　segmental arch
弓形拱涵　segmental arch culvert
公路路堤　highway embankment
供浆量　grout delivery
汞测压计　mercury gauge
拱脚推力　thrust at springer
拱矢　rise of arch
拱作用　arch action, arching
拱座压力　abutment pressure
共轭应力　conjugate stress
共价键　covalent bond
共振法　resonance method
共振频率　frequency at resonance
共振柱试验　resonant column test
共轴性　coaxiality
沟槽回填　trench backfill
沟蚀　gully erosion
构造地貌　structural landform
构造地震　tectonic earthquake
构造地质学　architectonic geology
构造节理　tectonic joint
构造结构面　tectonic structural surface(struc-

ture plane)
构造裂缝　tectonic fissure
构造面　constitutive surface
构造特征　tectonic feature
构造性地裂缝　tectonic ground crack
构造应力　tectonic stress
构造运动　tectonic movement
谷坊　check dam
孤立裂隙　isolated interstice
古滑坡　ancient rock slide, fossil landslide
古近纪　Paleogene period
古生代　Palaeozoic era
古土壤　oil soil, fossil soil
谷特曼法(化学加固)　Guttman process
骨架曲线　skeleton curve
骨料　aggregate
骨料灌浆混凝土　grouted-aggregate concrete
鼓起(路工)　blow-up
固壁泥浆　stabilizing fluid
固定地下水　fixed ground water
固定端　restrained end
固定环式固结仪　fixed-ring consolidometer
固定活塞式取样器　fixed piston sampler, stationary-piston sampler
固定锚杆　fixing bolt
固定式灰浆搅拌机　stationary mortar mixer
固化　cure, stiffening
固结　consolidation
固结(渗压)曲线　consolidation curve
固结百分率　percentage of consolidation
固结比　consolidation ratio
固结变形　consolidation deformation
固结不排水剪切试验　consolidated undrained shear test

固结不排水三轴试验	consolidated undrained triaxial test
固结不排水三轴压缩试验	consolidated-drained triaxial compression test
固结不排水试验	consolidated undrained test
固结不排水压缩试验	consolidated undrained compression test
固结沉淀作用	consolidation sedimentation
固结沉降	consolidation settlement
固结度	degree of consolidation, percent consolidation
固结灌浆	consolidation grouting, stabilizing grout
固结过程	process of consolidation
固结快剪试验	consolidated quick(direct) shear test, R-test
固结理论	theory of consolidation, consolidation theory
固结慢剪试验	consolidated slow shear test
固结排水	drainage by consolidation
固结排水三轴试验	consolidated drained triaxial test
固结排水试验	consolidated drained test
固结排水直剪试验	consolidated-drained direct shear test
固结破坏	consolidation failure
固结曲线	consolidation line
固结试验	oedometer test, consolidation test
固结速率	rate of consolidation
固结系数	coefficient of consolidation
固结压力	consolidation pressure
固结仪	consolidation apparatus, consolidometer, oedometer

固结仪三轴仪联合试验法	oedotriaxial test
固墙竖撑	dead shoring
固体力学	solid mechanism
固有刚度	self-stiffness
固有抗剪强度	intrinsic shear strength
固有抗剪强度包线	intrinsic shear strength curve
固有频率	natural frequency
固有特性	intrinsic property
固有振动	natural vibration
固有周期	natural period
刮板	screed board
刮刀	scraping cutter
刮土板	scraping blade
刮土地平机	scraping grader
拐点	inflection knee(point), knee
拐点法	inflection point method
观测井	observation borehole, recorder well
观测廊道	instrument gallery
管道	conduit
管段沉入	tube-immersing(sinking)
管段刚性接头	rigid joint between tube segments
管涵	pipe culvert
管井排水	tube well drainage
管棚法	tunneling with pilot pipes
管式排水	pipe drainage
管式取样器	tube sampler
管套	pipe sleeve, pipe casing
管涌	piping
管涌比	piping ratio
管涌破坏	failure by piping
管涌侵蚀	piping erosion
管柱基础	colonnade foundation, tubular colonnade foundation, tube caisson

foundation
管柱基础施工法 colonnade foundation process
管柱桩 caisson pile
管桩 pipe pile, tubular pile
管状基础 tubular foundation
贯入技术 penetration technique
贯入深度 depth of penetration
贯入试验 penetration test
贯入试验锤击 blow of penetration test
贯入速率 rate of penetration, penetration rate
贯入针 penetration needle
贯入阻力 penetration resistance
惯性 inertia
惯性矩 moment of inertia
惯性力 inertia force
惯性主轴 principal axes of inertia
灌浆 grouting
灌浆半径 grouting radius
灌浆泵 grouting pump
灌浆材料 grouting material, injection material
灌浆参数 grouting parameter
灌浆成果 grouting result
灌浆段 grouting stage
灌浆盖板 grout cap
灌浆工程 grouting project
灌浆管 grout injection pipe, grout pipe, injection pipe
灌浆盒 grout box
灌浆机 grouting machine
灌浆结石 set grout
灌浆截水墙 grouted cut-off wall
灌浆孔 grout hole
灌浆孔塞 packer
灌浆廊道 grout gallery

灌浆乱石　grouted riprap
灌浆锚杆　grouted anchor, grouted bolt
灌浆凝聚力　grout cohesion
灌浆区　grouted area
灌浆日报表　daily grouting report
灌浆试验　grouting test
灌浆顺序　sequence of grouting
灌浆土工布护排　grout-filled fabric mat
灌浆帷幕　grout curtain
灌浆效果　grout effectiveness
灌浆压力　grouting pressure, injection pressure
灌浆终结　termination of grouting
灌浆钻孔　drilled grout hole
灌浆作业　grouting operation
灌泥浆　mud grouting
灌入　injection
灌入量　volume injected
灌砂法　sand cone method
灌水下沉　sinking by water filling
灌注流量　injection flow
灌注桩　cast-in situ pile, cast-in-place pile, filled pile
光面爆破　smooth surface blasting
光面岩石　rock block with preconditioned smooth surface
广义毛管势　matric potential
广义条分法　generalized procedure of slices
龟裂黏土　fissured clay
规范沉降计算法　settlement calculation by specification
硅粉混凝土　silica fume concrete
硅化法　silicification
硅化灌浆　silicification grouting, solidification

grouting
硅石　silica
硅酸盐注浆　silicate injection
硅土水泥砂浆　silica cement
硅藻土　diatomaceous earth, diatomite, randanite
硅质软泥　silicareous ooze
轨束梁　rail beam
轨枕　rail bearing
过饱和　oversaturation
过大沉降　excessive settlement
过度碾压　over-rolling
过度压实　overcompaction
过量抽水　overpumping
过滤　filtration
过滤井　screened well, strainer well
过敏性黏土　quick clay
过筛砂　screened sand, sifted sand

H

哈佛小型碾实试验　Harvard miniature compaction test
海岸沉积　coast deposit
海岸地下水　coastal ground water
海岸阶地　coast terrace
海岸砂　coast sand
海岸淤填土　beach fill
海拔高度　elevation above sea level
海成岩石　marine rock
海堤　seawall
海底取样器　Cambridge sampler, Kullenberg sampler

海底隧道　undersea tunnel
海底土　submarine soil
海底淤泥　ooze
海底钻探　offshore boring
海积土　marine soil
海蚀台地　abrasion terrace
海滩　sea beach
海滩砂　beach sand
海湾三角洲　bay delta
海相沉积　marine deposit
海相黏土　marine clay
海啸　tsunami
海洋沉积　oceanic deposit
海洋工程学　ocean engineering
海洋结构物　ocean structure
海洋取样　marine sampling
海洋土　ocean soil
海洋土力学　marine soil mechanics
海洋岩土工程学　marine geotechnology
含冰率　ice content
含大量沙砾的河流　debris laden stream
含壳砾石　coated gravel
含气量　gas content
含气率　air content, percentage air voids
含沙量　sediment concentration
含沙重量比浓度　sediment concentration as percentage by weight
含水比　water content ratio
含水层　water bearing bed (layer), water-bearing stratum, aquifer
含水层等压线　isopiestic line of aquifer
含水当量　moisture equivalent
含水量盒　moisture can
含水量-密度核子测定仪　moisture-density nu-

clear gauge
含水率试验 water content test
含水率 moisture content, water content
含酸量 acidic content
含铁红壤 ferruginous lateritic soil
含盐地下水 salt ground water
含盐类岩石 salt-bearing rock
含盐量 salt content
涵洞 culvert
寒武纪 Cambrian period
夯锤 pounder, ram
夯击面积 tamping area(range)
夯击能 tamping energy
夯击碾 impaction roller
夯坑 crater
夯实 tamping, compaction by tamping
夯实回填土 tamped backfill
夯实混凝土 rammed concrete
夯实扩底混凝土桩 compacted expanded base concrete pile
夯实黏土 rammed clay
夯实水泥土桩法 rammed-cement-soil pile
夯实土桩 rammed-soil pile
夯土机 tamper, rammer compactor
夯土墙 rammed earth wall
航空地质 aerial geology
航空照片判读 airphoto interpretation
耗浆量 grout consumption, grout take
合金钻头 alloy drill bit
合力 resultant(resulting force)
合力曲线 resulting force curve
和易性 workability
河岸沉陷 bank settlement
河岸稳定 stream bank stabilization

河成阶地　fluvial terrace
河床　river bed
河谷阶地　valley terrace
河海沉积　fluvio-marine deposit
河口沉积　estuary deposit, potamogenic deposit
河流槽蚀　stream fluting
河流沉积　fluvial deposit, river drift
河流冲积扇　river fan
河流冲积土　fluvial soil
河流冲蚀　fluvial erosion
河流冲刷　stream erosion
河流动力学　river dynamics
河流工程学　river engineering
河流合流作用　stream abstraction
河流流域　river basin
河流能量　stream energy
河流泥沙　river sediment
河流泥沙量　stream load
河流侵蚀　river erosion
河流作用　stream action
河漫滩　alluvial flat, valley flat
核废料处理　nuclear waste disposal
核实测量　confirmatory measurement
核子密度仪　nuclear densometer
核子土壤湿度计　nuclear soil moisture meter
荷兰三轴仪　Dutch cell
荷兰式三轴试验　cell test
荷兰式圆锥触探仪　Dutch cone penetrometer
荷兰锥贯入仪　Ducth cone penetrometer
荷载-沉降曲线　load-settlement curve
荷载传递　loading transfer
荷载传递法　load transfer method
荷载传递机理　load transfer mechanism
荷载传感器　loading sensing cell

荷载大小　magnitude of load
荷载代表值　representative value of load
荷载分项系数　partial factor for load
荷载峰值　load peak
荷载函数　loading function
荷载架　loading frame
荷载阶段　load stage
荷载历时　loading duration
荷载历史　loading history
荷载挠度曲线　load deflection curve
荷载倾角　inclination of load
荷载倾斜因数　inclination correction factors
荷载势能　potential energy of loads
荷载-位移曲线　load-displacement curve
荷载系数　load coefficient, load factor
荷载系数法　load factor method
荷载效应组合　load effect combination
荷载增量　load increment
荷载增量比　load increment ratio
荷载增量法　increment(al) load method
荷载组合　load combination
盒式剪切仪　box shear apparatus
盒作用因数　cell action factor
褐煤　lignite
褐铁矿　ferrohydrite, hydrosiderite, limonite, ortstein
褐云母　anomite
黑钙土　chernozem soil
黑棉土　black cotton soil
黑土　black soil
黑云母　biotite
恒定流　steady flow
恒力幅激振　constant-force-amplitude excitation

恒体积试验	constant volume test
恒压装置	constant pressure device
恒载	dead load
横断面图	cross section profile, section diagram
横断面桩	cross section line stake
横观各向同性体	lateral isotropy
横节理	orthogonal joint
横面渗透性	cross-plane permeability
横向荷载	lateral load
横向力	lateral force, sideways force
横向透水性	transverse permeability
横向载荷试验	lateral load test
横向支撑	lateral support
烘干	adustion
烘干土	oven-dried soil
烘箱	dry oven, drying oven
红黏土	adamic earth, red clay
红土	laterite
红土化作用	laterization
红土浆	red ochre grout
红外光分析	infrared analysis
红外探测	infrared detection
宏观结构	macrostructure
宏观组构	macrofabric
洪积层	diluvion(diluvium), pluvial, proluvium
洪积扇	diluvial fan
洪积土	diluvial soil
洪积物	drifted material, flood deposit
虹吸管	siphon pipe
后端	rear-end
后拉锚定系统	anchored tied-back shoring system
后坡	back slope

后屈曲平衡状态　post-buckling equilibrium state
后屈曲强度　post-buckling strength
后效　after-working
后支撑　rear support
厚壁取土器　thick wall sampler
厚木板桩　plank pile
湖成泥灰岩　lake marl
湖泥　lake mud
湖相沉积　lacustrine deposits
虎克定律　Hooka's law
虎克模型　Hookean model
虎克体　Hookean body
互层　alternating beds, interbedded strata, interbedding
护岸　bank protection, revetment bank
护岸工程　bank protection work
护壁泥浆　slurry coat, wall-protecting slurry
护壁桩　retaining pile
护堤　revetment dike
护堤桩　embankment pile
护脚　landslide prevention skirting
护脚石　apron stone
护坡　slope protection, revetment
护坡的上下限　upper-lower limit of slope protection
护坡工程　slope protection works
护墙　guard wall, revetment wall
护坦　apron
护坦板　apron slab
护坦笼框　apron crib
护筒　pile casing
护桩　fender pile, pile fender
花岗石　granite, moorstone
滑动打桩机　skid pile-driver

滑动带	slip zone
滑动角	angle of slide
滑动面	plane of sliding, slip surface
滑动稳定性	sliding stability
滑动线	line of sliding, slip line
滑动压力	hovering pressure
滑坡	eboulement, landslide, slope failure
滑坡变形	slide deformation
滑坡加固	stabilization of landside
滑坡监测	monitoring of landslide
滑坡体	landslide mass
滑坡影响	landslide effects
滑坡整治	landslide correction
滑石	talc
滑楔试算法	trial wedge method
滑移力	skid force
化学板排水	chemical board drain
化学沉积	chemical deposit, chemical precipitation
化学废料处理	chemical waste disposal
化学分解	chemical decomposition
化学风化	chemical weathering
化学灌浆	chemical grouting
化学加固	chemical stabilization
化学胶合法	chemical bonding
化学试剂	chemical reagents
化学添加剂	chemical additive
化学稳定性	chemical stability
化学相互作用	chemical interaction
化学注浆法	chemical injection process
划痕测硬法	scraping hardness test
还土	earth backing
环斑花岗石	Rapakivi granite
环刀	cutting ring, edged ring

环剪　bromhead ring shear
环剪试验　ring shear test
环境水文地质学　environmental hydrogeology
环境岩土工程学　environmental geotechnics
环向应力　circumferential stress, hoop stress
环形沉井法　roundabout open caisson method
环形基础　ring(shaped)foundation
环形钻头　annular bit
环氧基粘合　epoxy bonding
环氧树脂　epoxy resin
缓冲桩　cushion pile
缓凝剂　set-retarding admixture
缓凝作用　set-retarding
换砂法　sand replacement method
换水法(测密度)　water displacement method, water replacement method
换算截面　transformed section
换算截面法　transformed section method
荒漠土　desert soil
黄铁矿　pyrite
黄土　loess
黄土高原　loess plateau
黄土结核　loess concretion
黄土类土　loessial soils
黄土湿陷　loess collapsing
黄土湿陷试验　collapsibility test of loess, loess collapsibility test
黄土湿陷性　collapsibility of loess
黄土状土　loess-like soil
灰泥岩　calcilutite(calcilutyte)
灰泥注浆　injecting grout
灰土　lime treated soil
灰土垫层　lime-soil cushion
灰土夯实　lime earth rammed, rammed lime

earth
灰土基础　lime-soil foundation
灰土挤密桩　lime-soil compaction pile
灰土挤密桩加固法　soil compaction by lime-soil pile
灰土井柱　lime-soil column
灰土桩　lime-soil pile
灰质黏土　lime clay
灰质软泥　caustic lime mud
恢复力　restoring force
恢复力模型　restoring model
恢复力特性　restoring force character
恢复系数　coefficient of restitution
辉长石　gabbro
辉石　pyroxene
回弹　rebound, resilience
回弹变形　radius deformation, rebound deformation
回弹法　rebound method
回弹模量　modulus of resilience, rebound modulus
回弹曲线　rebound curve
回弹系数　coefficient of resilience
回弹仪　rebound apparatus, resiliometer
回弹应变　radius strain
回弹应力　radius stress
回弹指数　rebound index, resilience index, swelling index, expansion index
回风井　upcast shaft
回灌　recharging
回灌法　recharge method
回灌井　recharge well, inverted well
回流阀　backflow valve
回流管　backflow connection

回声测深　acoustic depth sounding
回声测深仪　acoustic sounder, fathometer, echograph
回水　backwater
回填　back filling, refilling
回填灌浆　backfill grouting
回填夯　backfill tamper
回填密度　back fill density, backfill density
回填土　back fill
回跳打桩机　rebound pendulum machine
回转钻孔灌注桩　rotatory boring cast-in-place pile
汇水面积　catchment area
绘制草图　sketch drawing
混合沉积岩　mixed sedimentary rock
混合体力学　mechanics of mixture
混合桩复合管　composite pile
混凝土板桩　concrete sheet piling
混凝土垫层　concrete cushion
混凝土骨料　aggregate for concrete
混凝土管桩　concrete tubular pile
混凝土护层　blinding concrete
混凝土基础　concrete foundation
混凝土截水槽　concrete key trench
混凝土抗渗性　seepage resistance of concrete
混凝土离析　segregation of concrete
混凝土锚桩　concrete deadman
混凝土面板堆石坝　concrete face rockfill dam
混凝土木桩　concrete-timber pile
混凝土碎石　concrete rubble
混凝土重力式钻采平台　concrete gravity platform
混凝土桩　concrete pile
混凝土钻探抽样测试　random coring

混杂土　blended soil
活动层　active zone
活动性指数　activity index
活断层　active fault
活断裂　active fracture
活荷载　live load
活荷载产生的土压力　earth pressure due to live load
活化剂　active agent
活塞泵　displacement pump
活塞取芯器　piston corer(sampler)
活性　activity
活性比　activity ratio
活性层　active layer
活性加固剂　active stabilizer
活性黏土　active clay
活性膨润土　activated bentonite
活性水　active water
活性土　active soil
火成岩　igneous rock
火山地震　volcanic earthquake
火山灰　volcanic ash
火山灰水泥　pozzolan cement
火山泥流　lahar
火山岩　volcanic rock
火山岩风化土　volcanic saprolite

J

击穿阻力　puncture resistance
击入式取土器　drive sampler
击实功　compaction effort
击实曲线　moisture-density curve, compaction

curve
击实试验　moisture-density test, compaction test, compacting test
击实筒　compaction mold
击实仪　compaction device
击数　blow count
机场分类法　airfield classification system
机动洛阳铲　power-driven Luoyang spoon
机理　mechanism
机器基础　machine foundation
机械铲　power shovel
机械风化　mechanical weathering
机械阻抗　mechanical impedance
积水现象　waterlogging
基本烈度　basic intensity
基槽　foundation trench
基槽支撑　shoring of trench
基层　foundation course
基础　footing, foundation, ground work
基础变形　foundation deformation
基础标高　level of foundation
基础布置平面图　foundation layout plan
基础沉陷　yield of foundation
基础承台　foundation platform
基础底板　foundation slab, foundation mat
基础底面压力　foundation pressure
基础垫层　foundation pad
基础放样　setting out of foundation
基础分析　foundation analysis
基础刚性角　load distribution angle
基础隔振(震)　foundation isolation, vibration isolation of foundation
基础工程　foundation engineering
基础沟　foundation ditch

基础灌浆加固　foundation by means of injecting cement
基础荷载　foundation load
基础换托　foundation underpinning
基础加固　foundation improvement
基础降水　foundation dewatering
基础结构　foundation structure
基础襟边　offset of foundation
基础连系梁　linking beam of foundation
基础梁　footing beam, foundation beam
基础露出　uncovering of foundation
基础埋深　depth of foundation
基础埋置深度　foundation embedment
基础锚固　foundation anchoring
基础排水孔　foundation drain hole
基础破坏　foundation failure
基础墙　foundation wall
基础切向冻胀稳定性　tangential frost-heaving stability of foundation
基础施工　foundation construction
基础施工技术　foundation practice
基础台阶宽高比的容许值　allowable value of width-height of foundation steps
基础抬升　jacking of foundation
基础特征　foundation characteristics
基础托换　underpinning
基础托换法　underpinning method
基础位移　foundation displacement
基础稳定性　stability of foundation soil
基础线　foundation line
基础形状系数　shape factor of foundation
基础振动　foundation vibration
基础支撑　shoring of foundation

基础置换　replacement of foundation
基础桩　foundation pile
基础自振频率　natural frequency of foundation
基础最小埋深　minimum embedded depth of foundation
基床　foundation bed
基床系数法　subgrade coefficient method
基底　foundation base, basement
基底附加压力　net contact (foundation) pressure
基底隆起　bottom heave
基底摩擦角　angle of base friction
基底倾斜因素　base tilt factor
基底压力　pressure on foundation soil, gross loading intensity
基底标高　foundation level
基底灌浆桩　grouted base pile
基脚　foundation footing
基坑　foundation pit
基坑侧壁　side of foundation pit
基坑底隆起　basal heave
基坑底隆胀　heaving of the bottom
基坑回弹系数　coefficient of foundation ditch's rebound
基坑排水　foundation pit drainage
基坑支撑　rafter timbering
基坑支护　retaining and protection of building foundation excavation
基坑支架　rafter set
基石　bedrock, foundation stone, sill
基岩　base rock, bed rock
基岩地震动　seismic bedrock motion
基岩地质图　bedrock map
基岩加速度　bedrock acceleration

基岩埋深图　map of burial depth of bedrock
基于可靠度的优化设计　reliability-based optimum design
基质吸力　matrix suction
基准面　datum plane
基准线　base line
基准钻孔　key well
箕斗提升机　skip winder
激震装置　impulse seismic device
吉布森地基模型　Gibbson soil
级配　grading, gradation
级配不良的　poor graded
级配不良集料　poor graded aggregate
级配反滤层　graded filter
级配均匀土　uniform soil
级配良好土　well-graded soil
级配土　gradated soil
级配系数　grading factor
极惯性矩　polar moment of inertia
极化剪切波速　polarized shear wave velocity
极化性　polarizability
极力　polar force
极限承载力　ultimate bearing capacity
极限承载阻抗　ultimate bearing resistance
极限分析　limit analysis, ultimate analysis
极限荷载　limiting load, collapse load, ultimate load
极限荷载法　ultimate load method
极限荷载设计法　ultimate load design method
极限抗拔力　ultimate pullout capacity
极限抗剪强度　ultimate shearing strength
极限抗拉强度　ultimate tensile strength
极限抗压强度　ultimate compression(compressive) strength

极限拉应变　ultimate tensile strain
极限摩擦角　ultimate friction angle
极限平衡法　limit equilibrium method
极限平衡理论　ultimate balance theory
极限平衡力学　limiting equilibrium mechanics
极限平衡状态　state of limit equilibrium
极限破坏荷载　ultimate breaking load
极限强度　limit strength, strength ceiling, ultimate strength
极限强度设计　ultimate strength design
极限设计　limit design
极限弯曲力矩　ultimate bending moment
极限弯曲能力　ultimate flexural capacity
极限弯曲强度　ultimate bending strength
极限压力　limit pressure, ultimate pressure
极限压应变　ultimate compressive strain
极限应力　ultimate stress
极限应力状态　ultimate stressed state
极限优化设计　optimal limit design
极限粘结强度　ultimate bond strength
极限值　limit value, ultimate value
极限状态　limit state, ultimate limit state
极限状态设计法　limit state design method
极限阻力　limiting resistance
极性分子　polar molecule
极值点失稳　limiting point instability
集合体　aggregation
集料最大粒径　maximum aggregate size
集水沟　collector drain, gulley
集水井　collector well
集中荷载（点荷载）　concentrated load, point load
集中力　concentrated force
集中因素　concentration factor

集中指数　concentration index
集总参数法　lumped parameter method
集总质量　lumped mass
几何各向同性　geometrical isotropy
几何近似　geometrical approximation
几何平均　geometrical mean
几何相似法　geometrical similarity method
几何学水头　geometric head
几何阻尼　geometrical damping
挤浆砌砖法　shove joint in brickwork
挤密灌浆　compaction grouting
挤密加固　stabilization by densification
挤密喷浆法　compaction grouting method
挤密砂桩　compaction sand pile, densification by sand pile
挤密砂桩加固法　sand compaction pile method
挤密碎石桩加固法　soil compaction by stone pile
挤密土桩法　soil compaction pile method
挤密桩　compacted column, compaction pile
挤密桩法　compaction pile method
挤压破碎带　compress-crushed zone
挤压实验　squeeze test
挤淤法　displacement method
计算荷载　assumed load
计算自重湿陷量　wet-subsidence due to overburden
计算自重性湿陷量(Δ_{zs})　calculated self-weight collapse
季节调节水库　seasonal reservoir
季节冻土　seasonally frozen soil
季节性变化系数　seasonal variation factor
季节性河流　seasonal stream
季节性荷载　seasonal load

季节性枯竭	seasonal depletion
加工膨润土	processed bentonite
加工软化	work softening
加工石料	processed rock
加工硬化	work hardening
加工硬化定律	strain harding law
加固板	reinforcement plate
加固层	reinforced layer
加固的路肩	sealed shoulder
加固堤	reinforcing dam
加固方案	layout of reinforcement
加固结构	reinforced structure
加固支撑	reinforcing stay
加固桩	consolidating pile
加筋挡土墙	reinforced fill wall
加筋法	reinforcement method
加筋路堤	reinforced embankment
加筋土	reinforced earth
加筋土挡墙	reinforced soil wall
加筋土挡土墙	reinforced(earth)retaining wall
加筋土桥台	reinforced earth abutment
加密	densification
加强板	reinforcing plate
加强筋	reinforced rib
加强肋	reinforcing rib
加强梁	reinforcing girder
加权平均值	weighted mean value
加水爆破法	hydro-blasting
加速固结	accelerated consolidation
加载触探仪	weight sounding
加载方式	loading method
加载路径	path of loading
加载排水	drainage by surcharge
加载强度	loading intensity

加载与卸载　loading and unloading
加州承载比(美国)　California Bearing Ratio (CBR)
加州承载力比值(美国)　CBR value
夹层　intercalation
夹层结构　sandwich structure
夹砂砾石　hoggin
夹桩力　pile locking force
架空砾石层　open work gravel
架状硅酸盐结构　framework silicate structure
假设承载力　presumptive bearing capacity
尖头桩　pointed pile
坚固度　competent degree of rock
坚固完整岩石　sound unaltered rock
坚石　solid rock
间层　interstratified bed
间断级配　gap grading
间隔挡板　open sheeting
间隔挡土板　openlagging
间接排水　indirect drainage
间隙　interstice
间隙水　interstitial water
间歇荷载　intermittent load
监测　monitoring
减饱和作用　desaturation
减压井　pressure relief well, relief well, bleeder well
减压岩石　decompressed rock
减振器　buffer
剪力板　shear plate
剪力传递装置　shear-transfer device
剪力开裂荷载　shear cracking load
剪力流　shear flow
剪力墙　shear wall

剪力图　shear force diagram
剪力滞后　shear lag
剪力滞后效应　shear lag effect
剪力中心　shear center
剪裂　shear fracture
剪摩擦　shear friction
剪切　shear, shearing
剪切闭锁　shear locking
剪切变形　shear deformation
剪切变形模量　modulus of shear deformation
剪切波　distortional wave, S-wave
剪切承载力　shear capacity
剪切带　shear band
剪切弹性模量　modulus of elasticity in shear, shear elasticity
剪切断面　shearing section
剪切幅度　shear-range
剪切荷载　shear load
剪切盒试验　shear box test
剪切滑动　shear slip
剪切滑塌　shear slide
剪切回弹模量　modulus of shear resilience
剪切极限　shearing limit
剪切角　angle of shear (shearing resistance, shearing strength)
剪切节理　shear joint
剪切理论　shear theory
剪切力　shearing force
剪切力矩　shearing moment
剪切力图　shearing force diagram
剪切裂缝　sheared crack, shear crack
剪切裂纹　shear fissure
剪切面　plane of shear, shear plane
剪切面积　shear area

剪切模量　shear modulus
剪切摩阻理论　shear-friction theory
剪切黏度　shear viscosity
剪切破坏　fail in shear, shear failure, shearing failure
剪切破坏安全系数　safety factor against shear failure
剪切破坏模式　shear failure model
剪切强度　shearing strength
剪切强度条件　condition of shear strength
剪切区　shear zone
剪切试验　shear test, shearing test
剪切型变形　shearing type deformation
剪切仪　shear apparatus
剪切应变速率　shear strain rate
剪切震动　shearing vibration
剪式钻臂　shear type drill boom
剪缩回胀性　compression and back expansion
剪缩剪胀性　shear contraction and dilation
剪塌　shear distortion
剪压比　shear compression ratio
剪压破坏　shear compression failure
剪应变　shearing strain, shear strain
剪应变速率　rate of shear strain
剪应力　shear stress, shearing stress
剪应力差法　shear stress difference method
剪应力轨迹　shearing stress trajectory
剪胀(性)　dilatancy, dilatation, dilation
剪胀方程　dilatancy equation
剪胀角　angle of dilatancy
剪胀性土　dilative soil
检测器　monitor
检查井　inspection well, manhole
检查孔　check hole, inspection borehole

检验荷载　proof load
简谐振动　harmonic vibration
碱度　alkalinity
碱性水　alkaline water
碱性土　alkali(ne) soil
建筑地基基础设计规范　foundation design code for building
建筑放线　setting out of building
剑桥模型　Cambridge model
剑桥旁压仪　Camkometer
渐进破坏　progressive failure
渐进性沉降　progressive settlement
渐进性滑动　progressive sliding
鉴定试验　qualification test
键结　linkage
键力　bonding force
浆料输送　slurry transport
浆砌拱圈　mortar bound arch
浆液　grout
浆液搅拌机　grout agitator
浆液扩散半径　radius of grouting spread
浆液排放阀　grout discharge valve
浆液配比　grout mix
浆液配方　grout formulation
浆液配合比设计　grout mix design
浆状炸药　slurry explosives
降低地下水位　dewatering, groundwater lowering
降落跑道　landing runway
降水法　dewatering method
降水工程　dewatering project
降水量　precipitation
降水漏斗　depression-cone
降水掏土纠偏　tilting correction by undercut-

ting and dewatering
降水系统　dewatering system
降雨量分布图　isohyetal map
交变荷载　alternating load
交变应变　alternating strain
交变应力　alternating stress
交叉各向异性土　cross-anisotropic soil
交叉排水渠　drainage crossings
交错层理　cross-stratification
交错层面　cross bedding
交点　intersection point
交角　intersection angle
交面　intersection plane
交通隧洞　access tunnel
交线　intersection line
浇灌　pouring
浇灌混凝土　concreting, pour concrete
浇筑顺序　sequence of pours
胶合　bonding
胶结　bond
胶结剂　cementing agent
胶结黏土　bond clay
胶结土　cemented soil
胶结作用　adhesion action, cementation
胶粒　colloidal particle, micellae
胶粒粒组　colloid fraction
胶体悬液　colloidal grout, colloidal suspension
胶粘剂　bonding agent
角点法　corner-points method
角砾　angular pebbles
角砾岩　breccia
角粒　angular grain
角挠曲　angular distortion
角闪石　hornblende, keraphyllkite

角位移　angular displacement
铰接　joint with hinge
搅拌　agitation
搅拌机　stirring machine
搅拌机械　rabbling mechanism
搅拌器　agitator
阶地　terrace
阶梯形地基　stepped foundation
接触灌浆　contact grouting, surface grouting
接触角　angle of contact
接触面冲蚀　contact erosion, contact scouring
接触面积　contact area
接触面强度　interface strength
接触压力　contact pressure
接触应力传感器　contact stress transducer
接缝灌浆　joint grouting
接缝止水　joint seam
接合剂　adhesive
接头施工　joint construction
接桩　pile extension
节点力　nodal force
节点位移　nodal displacement
节理　joint
节理玫瑰图　rose diagram of joints
节理面　joint plane, divisional plane
节理频率　joint frequency
节理岩石力学　mechanics of jointed rock
节理岩体　jointed rock
节理组　joint set
结构表面系数　specific surface coefficient of structure
结构的可靠性　reliability of structure
结构面　structural plane
结构面的剪切变形　shear deformation of struc-

ture plane
结构面抗剪强度　shear strength of structure plane
结构特征　structural characteristic
结构体　structural block
结合层　binder course
结合膜　contractile skin
结合水　combined water, held water, bound water
结核　accretion
结晶水　crystal(lization) water, water of crystallization
结晶岩　crystalline rock
结壳　incrustation
截流　stream closure
截流井　intercepting well
截流水力学　stream closure hydraulics
截留盲井　blind catch basin
截面　cross section
截面尺寸　sectional dimension
截面尺寸变量　sectional sizing variables
截面惯性矩　moment of inertia of cross-section
截面积　sectional area
截面剪切变形的形状影响系数　shape influence coefficient of section shear deformation
截面矩　section moment
截面模量　modulus of section
截面收缩率　reduction of area
截面塑性系数　plastic factor of section
截面图　sectional drawing
截面形状系数　shape factor of section
截面主惯性轴　inertia principal axis of area
截水沟　interceptor drain

截水墙　cut off wall, diaphragm wall
解冻　defrosting
介质传导率　conductivity of medium
界面剪切强度　interfacial shear strength
界面力　interfacial force
界面速度　interface speed
界面粘结强度　interfacial bond strength
界面张力　interfacial tension
界限约束　side constraints
金刚石钻头　diamond crown bit
金刚石钻头钻探　diamond drilling
金属键　metallic bond
紧密堆积理论　theory of high packing
近岸　nearshore
近海沉积　offshore deposit
近海阶地　offshore terrace
近海结构物　nearshore structure, offshore structure
近海台地　offshore platform
近海土力学　offshore soil mechanics
近海桩　offshore piles
近海钻井平台　offshore well drilling platform
浸没密度　buoyancy density
浸没土的密度　density of submerged soil
浸润面　soaking surface, soaking line, wetting front
浸湿土　water logged soil
浸水　inundation
浸水试验　ponding test
浸渍试验　immersion test
经验常数　empirical constant
经验公式　empirical equation, empirical formula
经验值　empirical value

晶格　crystal lattice
晶格电荷　lattice charge
晶格结构　crystal structure, lattice structure
晶棱　crystal edge
晶体不完整　crystal imperfection
井　well
井壁　wall of a well
井壁破裂　rupture of shaft lining
井壁稳定　stable borehole
井壁岩芯　sidewall core
井出水量　capacity of well, yield of well
井的渗流面积　seepage area of well
井的吸收容量　inverted capacity of well
井的有效半径　effective well radius
井底车场　shaft station
井点抽水　pumping from well point
井点法排水　well point method
井点管　riser pipe
井点降水　well point dewatering
井点排水　drainage by well point, well point
井点排水系统　well point pumping system
井点系统　well point system
井径测量　hole diameter measurement
井滤网　well screen
井内静水位　rest water level
井群　gang of wells
井筒基础　well foundation
井阻　well resistance
颈缩　necking
颈缩现象　necking phenomena
颈缩应变　striction strain
净荷载强度　net loading intensity
净容许承载力　allowable net bearing pressure
净水泥浆　neat cement grout

径流 runoff
径流计算公式 runoff computing formula
径流区 runoff area
径流系数 runoff coefficient
径切面 radial section
径向扁千斤顶法 radial flat jack technique
径向变形 radial deformation
径向分布函数 radial distribution function
径向固结 radial consolidation
径向固结系数 coefficient of consolidation for radial flow
径向剪切 radial shear
径向抗力 radial-thrust force
径向抗压强度 radial crushing strength
径向力 radial force
径向裂缝 radial fissure
径向裂纹 radial crack
径向流动 radial flow
径向排水 radial drainage
径向矢量 radial vector
径向速度 radial velocity
径向位移 radial displacement
径向压应力 radial compressive stress
径向应变 radial strain
径向应力 radial stress
径向张力 radial tension
径向正应力 radial normal stress
径向支撑 radial bracing, radial support
径向主应力 radial principal stress
静定分析法 static analysis
静荷载 static load
静力初探试验 static penetration test
静力触探 static cone sounding, static sounding

静力触探试验 static cone penetration test
静力触探探头阻力 static point resistance
静力弹塑性分析 pushover analysis
静力摩擦 static friction
静力试验 static test
静力压拔桩机 static pile pressure-extract machine
静力压桩 silent piling, static pile loading
静力压桩机 pile jacket
静力载荷试验 bearing test
静水上托力 hydrostatic uplift
静水位 static water level
静水位移监测计 hydrostatic profile gauge
静水压 static head
静水压力 hydrostatic pressure, static water pressure
静水压力分布 hydrostatic distribution
静水压力线 hydrostatic line
静水应力状态 hydrostatic state of stress
静态试验载荷 static test load
静压力 static pressure
静载荷试验 static load test
静止土压力 at rest pressure, earth pressure at rest
静止土压力系数 coefficient of earth pressure at rest
旧石器时代 chellean age
就地灌注地下连续墙 cast-in-place diaphragm wall
就地灌注桩 driven cast-in-place pile, in-situ pile
局部侧摩阻 local side friction
局部承压面积 local bearing area
局部冲刷 local erosion, partial erosion

局部剪切破坏	local shear failure
局部开挖	partial excavation
局部气压盾构	shield with partial air pressure
局部气压法	local pneumatic process
局部屈曲	local buckling
局部受压承载力	local bearing capacity
局部弯曲	local flexure
局部硬化	local hardening
矩形单元	rectangular element
矩形盾构	rectangular shield
矩形基础	rectangular foundation
矩形截面梁	rectangular cross section beam
矩形梁	rectangular girder
巨粒土	over coarse-grained soil
巨石块	derrick stone
拒浆标准	refusal criterion
拒浆流量	refusal flow rate
拒浆压力	refusal pressure
聚苯乙烯发泡材料	expanded polystyrene (EPS)
聚合力	aggregation force
聚合物加固	polymer stabilization
绝对标高	absolute altitude
绝对等高线	absolute contour
绝对高程	absolute elevation
绝对控制点	absolute control point
绝对密度	absolute density
绝对年代	absolute chronology
绝对湿度	absolute humidity
绝对值	absolute value
绝对坐标	absolute coordinate
绝热压缩	adiabatic compression
均布荷载	uniformly distributed load
均布应力	uniform stress

均衡异常　isostatic anomaly
均一粒径骨料　single sized aggregate
均匀沉降　uniform settlement
均匀级配　narrow gradation, uniform grading, uniformly graded
均匀粒径骨料　uniform size aggregate
均匀系数　uniformity coefficient
均匀压实　uniform compaction
均质层　homogeneous stratum
均质土　homogeneous soil
竣工地面高程　finished ground level
竣工检验　test on completion

K

喀斯特(岩溶)　karst
喀斯特的　karstic
喀斯特地基加固　consolidation of Karst
喀斯特地貌　karst land feature
喀斯特含水层　karst aquifer
喀斯特溶槽　karst channels
喀斯特塌陷　karst collapse
卡尔逊应力计　Carlson stress meter
卡萨格兰德孔隙压力测头　Cassagrande dot
卡萨格兰德土分类法　Cassagrande's soil classification
卡萨格兰德液限仪　Cassagrande liquid limit apparatus
开尔文模型　Kelvin model
开沟犁　mole-plough
开管式柴油锤　open tube diesel hammer
开孔面积比　open area ratio
开口钢桩　open-end steel pile

开口管桩	open-end pipe pile
开口裂隙	open fissure
开裂荷载	load at first crack
开派斯模型	Kepes model
开挖断面	excavated section
开挖基线	base of excavation
开挖临界深度	critical depth of excavation
开挖深度	excavation depth
开挖线	lined of excavation
凯塞效应	kaiser effect
勘测	exploration survey, reconnaissance
勘察	investigation
勘察阶段	investigation stage
勘探	exploration, exploratory, prospecting
勘探工作	exploration work
勘探阶段	exploration stage
勘探孔	prospect hole
勘探平洞	exploratory adit
勘探竖井	prospecting shaft
勘探线	exploratory line
勘探钻孔	exploratory bore-hole
抗拔力	pulling resistance, uplift resistance
抗拔强度	pull strength
抗拔区	resistant zone
抗拔试验	pull-out test
抗拔桩	uplift pile
抗侧弯能力	resistance to lateral bending
抗冲击强度	resistance to impact
抗刺强度	puncture strength
抗冻力	frost resistance
抗冻融强度	freeze-thaw resistance
抗冻融性	resistance to alternate freezing and thawing
抗滑安全	safety against sliding

抗滑安全系数　factor of safety against sliding
抗滑能力　skid-resisting capacity
抗滑稳定　stability against sliding
抗滑稳定性　resistance to sliding
抗滑系数　skid-resistance value
抗滑性　skid-resistance
抗滑移系数　slip coefficient
抗滑桩　anti-slide pile, slide-resistant pile
抗剪弹性模量　shearing modulus of elasticity
抗剪刚度　shear stiffness
抗剪加固　reinforcement for shearing
抗剪力　shearing resistance
抗剪能力　shear-bearing capacity
抗剪强度　shear strength
抗剪强度参数　shear strength parameter
抗剪强度总应力法　total stress analysis of shear strength, total stress approach of shear strength
抗剪切破坏安全　safety against shear failure
抗拉刚度　tensional rigidity, tensile strength, tension strength
抗力　resistance, resistant force
抗力分项系数　partial factor for resistance
抗力结构　resistant structure
抗磨损试验　abrasion resistance test
抗挠刚度　flexural rigidity
抗扭刚度　torsional rigidity
抗扭刚度系数　torsional stiffness factor
抗扭强度　torsional strength
抗劈力　resistance to splitting
抗疲劳强度　fatigue resistance
抗破坏能力　resistance to failure
抗切断强度　cutting resistance
抗倾覆安全　safety against overturning

抗倾覆安全系数 factor of safety against overturning
抗倾稳定 stability against overturning
抗渗剂 permeability reducing admixture
抗渗试验 impermeability test
抗渗性 impermeability, penetration resistant
抗弯截面模量 section modulus for a beam
抗压强度 compressive strength, compression strength, strength of compression
抗液化措施 liquefaction defence measures
抗液化强度，液化强度 liquefaction strength
抗粘性 resistance bond
抗震分析 seismic appraise
抗震基础 earthquake proof foundation
抗震加固措施 strengthening measure for earthquake resistance
抗震建筑 earthquake proof construction
抗震结构 earthquake resistant structure
抗震设计 aseismatic design, earthquake resistant design
抗震稳定性 anti-vibrating stability
抗抓拉强度 grab tensile strength
颗粒 grain
颗粒比表面积 specific surface area of particle
颗粒大小 particle size
颗粒度分析 grain size measurement
颗粒分布 particle distribution
颗粒分析 screen analysis, grain-size analysis, granulometry, particle size analysis
颗粒分析曲线 grain-size analysis curve
颗粒干燥器 pelleter drier
颗粒骨架 grain skeleton
颗粒级配 mechanical composition, distribution of grain size

颗粒级配曲线　particle grading curve
颗粒交联　particle cross-linking
颗粒粒度　coarseness of grading
颗粒密度　particle density
颗粒破碎率　grain breakage, percentage of particle breakage
颗粒强度　particle strength
颗粒细度　fineness of grain
颗粒形状　grain shape, particle shape
颗粒压碎强度　crushing strength of grains
颗粒直径　particle diameter
颗粒组合　particle association
壳体的应变和应力　strain and stress in shell
壳体基础　shell foundation
可测孔隙水压力的触探头　piezocone
可拆卸式钻头　detachable bit
可灌比　groutability ratio
可灌地基土　injectable soil
可灌性　groutability
可灌筑性　placeability
可交换离子　exchangeable ion
可靠性设计　reliability design
可靠性试验　reliability trials
可靠性系数　reliability coefficient
可靠指标　reliability index
可控应变试验法　strain-controlled testing method
可控应力试验法　stress-controlled testing method
可逆膨胀　reversible expansion
可破碎性　spallability
可溶性黏土　soluble clay
可溶性盐试验　soluble salt test
可行性设计　feasibility design
可行性研究　feasibility study
可行性研究勘查　reconnaissance exploration

可压缩性　coercibility
可钻性　drillability
克罗内克符号 δ　Kronecker delta
刻蚀作用　corrosion
坑　pit
坑槽探　pit exploitation
坑道　adit
坑底不稳定性　basal instability
坑碉展示图　developing chart of exploratory drift
坑木　timber
坑式托换　pit underpinning
空口　aperture
空气污染　air pollution
空气压缩机　air compressor
空气止回阀　air back valve
空气钻探　air flush drilling
空隙率　percentage of void
空心沉井基础　unfilled caisson foundation
空心圆柱体试验　hollow cylinder test
空眼　void
孔壁　borehole wall, hole wall
孔径分布分析　pore size distribution analysis
孔距　hole spacing
孔口标高　elevation of bore hole
孔口管　stand pipe
孔隙　pore space
孔隙比　void ratio
孔隙比与压力曲线　pore ratio-pressure curve
孔隙测定仪　void measurement apparatus
孔隙含量　void content
孔隙计　porosimeter
孔隙结构　pore structure
孔隙介质　pore medium

孔隙流体　pore fluid
孔隙率　porosity
孔隙气压力　pore air pressure
孔隙水　pore water, void water
孔隙水头　pore water head
孔隙水压力　pore water pressure
孔隙水压力计　pore-water pressure gauge
孔隙水压力监测　monitoring of pore-water pressure
孔隙水压力消散　pore water pressure dissipation
孔隙水压力仪　pore water pressure cell
孔隙水张力　pore water tension
孔隙吸力　pore suction
孔隙相　pore phase
孔隙压力　pore pressure, interstitial pressure
孔隙压力比　pore pressure ratio
孔隙压力量测　pore pressure measurement
孔隙压力系数　pore pressure coefficients, pore pressure parameters
孔隙压力系数 A　A-parameter
孔隙压力系数 B　B-parameter
孔隙压力消散　pore (water) pressure dissipation
孔下点式录像机　down-hole TV camera
孔压静力触探试验　piezocone test (CPTU)
孔眼面积百分数　percent open area
控制比降固结试验　consolidation test under controlled gradient
控制回填土　engineered fill
库尔曼法　Culmann's method
库尔曼图解法　Culmann construction
库尔曼线　Culmann line
库仑定律　Coulomb's law
库仑抗剪强度公式　Coulomb's equation for shear

库仑土破坏楔体　Coulomb's soil-failure prism
库仑土压力　Coulomb's earth pressure
库仑土压力理论　Coulomb's earth pressure theory
库仑准则　Coulomb criterion
库伦-纳维强度理论　Coulomb-Navier strength theory
垮塌角砾岩　ablation breccia
跨孔法　cross-hole method (shooting)
块石　block stone, rock block
块石混凝土　rubble concrete
块石基础　rubble foundation
块体剪切试验　block shear test
块体理论　block theory
块体锚固　block anchorage
块土　lump soil
块状试样　block sample, chunk sample
块状岩　massive rock
快剪试验　Q (quick direct shear) test, quick test, quick shear test
快凝水泥　accelerated cement
快速冲击夯实　rapid impact rammer
快速固结试验　quick consolidation test
快速掘进　speedy drivage
快速水泥　speed cement
快速压缩试验　fast compression test
宽级配土　broadly graded soils
宽裂缝　macrocrack
矿化水　mineralized water
矿脉　vein
矿物成分　mineralogical composition
矿物分析　mineralogical analysis

矿物骨架 mineral skeleton
矿物相 mineral phase
矿物学 mineralogy
矿渣 mine refuse, slag
矿渣波特兰水泥 slag Portland cement
矿渣集料 slag aggregate
矿渣膨胀水泥 slag expansive cement
矿渣水泥 slag cement
矿渣填料 slag fill
框格式挡土墙 crib retaining wall
框架式基础 frame foundation
框图 block diagram
奎尔波 Quer-wave
扩底 under reaming, enlarged base
扩底墩 belled pier, belled shaft
扩底基础 under reamed foundation, under-reamed foundation
扩底挖斗 belling bucket
扩底桩 belled pile, club-footed pile, pedestal pile, under-reamed pile
扩底钻孔桩 under reamed bored pile
扩孔机 reamer
扩孔理论 cavity expansion theory
扩孔钻 counter-bore, expanding auger, bit reamer
扩孔钻头 expanding bit, reamer bit
扩散层 diffusion layer
扩散渗流 diffuse seepage
扩散双电层 diffuse double layer
扩散速率 rate of diffusion
扩散性 diffusivity
扩眼钻头 enlarging bit
扩展基础 spread footing

L

垃圾填土　garbage dump
拉铲　dragline
拉长试验　elongation test
拉撑　tension brace
拉断破坏　tension failure
拉杆　tie rod
拉结石　bonded stone
拉力　tensile force
拉力测试　tensile test
拉力屈服点　tensile yield point
拉力桩　tension pile
拉裂　tension fracture
拉普拉斯算子　Laplacian operator
拉伸变形　stretch elongation, stretcher strain
拉伸模量　modulus of elongation, tensile modulus
拉伸破坏　fail in tension
拉伸曲线　tensile curve
拉伸试验　extension test
拉伸应变　tensile strain
拉伸粘结性能　tensile adhesion property
拉索桩　stay pile
拉特屈服准则　Lade yield criterion
拉应力　tensile stress
来达黏土　Leda clay
拦石　curbstone
蓝图　blue print
朗肯(朗金)土压力理论　Rankine's earth pressure theory
朗肯被动区　passive Rankine zone

朗肯状态　Rankine state
浪成阶地　wave built terrace
浪成台地　abraded platform
浪蚀　wave cutting
浪蚀台地　abrasion platform(terrace)
浪蚀岩石阶地　rock abrasion platform
老堆积土　paleo-deposited soil
老化　aging
老黄土　old loess
老黏土　aged clay, paleo clay
乐甫波　Love wave
雷蒙式桩　Raymond concrete pile
累积曲线　accumulation curve, cumulative curve
累积误差　accumulated error
累计筛余百分数　percentage of accumulated sieve residues
累年最冷月　secular coldest month
棱角度对比图　angularity chart
棱角度因素　angularity factor
棱镜罗盘　prismatic compass
离散化　discretization
离析　segregation
离析区　segregation zone
离析指数　segregation index
离心泵　centrifugal pump
离心法制混凝土打入桩　centrifugally-cast concrete driven pile
离心钢筋混凝土桩　centrifugal reinforced concrete pile
离心含水当量　centrifuge moisture equivalent
离心混凝土桩　centrifugal concrete pile
离心机　centrifuge
离心机模型　centrifuge modeling

离心力　centrifugal force
离心模型试验　centrifugal model test
离心试验　centrifuge test
离心预应力混凝土桩　centrifugal prestressed concrete pile
离心蒸压混凝土桩　centrifugal autoclaved concrete pile
离子键　ionic bonding
离子交换　ion exchange
离子交换量　ion exchange capacity
离子扩散　ion diffusion
离子流　ion flow
离子浓度　ion concentration
离子水化作用　ion hydration
离子替代　ion substitution
理想弹塑性变形模型　model of idealized elastic-plastic deformation
理想弹塑性模型　ideal elastoplastic model
理想弹性材料　perfect elastic material
理想刚塑性变形模型　model of idealized rigid-plastic deformation
理想颗粒级配曲线　ideal grading curve
理想塑性　perfect plasticity
理想塑性材料　perfect plastic material
理想液体　ideal liquid
力的分解　resolution of a force
力的平行四边形　parallelogram of forces
力学模型　model of mechanics
力学相似性　mechanical similarity criterion
力学性质　mechanical properties
立方体强度　cube strength
立方体试块　test cube
立井　vertical shaft
立井地压　rock pressure in vertical shaft

立井普通掘进　sinking vertical shaft by convention
立井特殊掘进　special method of shaft sinking
立面图　sectional elevation
立桩　soldier beam
沥青堵漏　asphaltic sealing
沥青灌浆　asphaltic grouting, bitumen grouting
沥青黏土　bituminous clay
沥青砂　sand asphalt, sand bitumen
砾砂　gravelly sand
砾石　gravel
砾石基础　gravel foundation
砾石排水沟　gravel drain, gravel-filled drain trench
砾石填充层　gravel pack
砾岩　conglomerate
砾质土　calculous soil, gravelly soil
砾质岩　psephyte
粒度　grainage, grainness, granularity
粒度分布曲线　size distribution curve
粒度分级　particle size classification
粒度级别　size division
粒度累积曲线　particle size accumulation curve
粒度系数　particle size factor, size factor
粒间斥力　interparticle repulsion
粒间键　interparticle bond
粒间胶结　intergranular cement
粒间摩擦　interparticle friction
粒间透水性　intergranular permeability
粒间压力　grain pressure, intergranular pressure
粒间引力　interparticle attractive force
粒间作用力　interparticle force

粒径	grain size
粒径分布	diameter distribution, grain-size distribution, particle size distribution
粒径分布曲线	grading curve, grain size distribution curve, grain-size distribution curve
粒径分级	grading fraction, grain diameter
粒径分析	gradation test, grading analysis, granularmetric analysis, mechanical analysis
粒径分析曲线	mechanical analysis curve
粒径级配	size grading
粒径界限	grain boundary
粒径累积曲线	grain size accumulation curve
粒径频率曲线	grain-size frequency curve
粒径曲线	grain-size curve
粒径组	fraction of partial size
粒径组成	grain size composition, granulometric composition
粒体力学	particulate mechanics
粒团组构	domain fabric
粒状土	granular soil
粒组	fraction
连拱墙衬砌	lining with continuous-arched sidewalls
连续底座	continuous footing
连续方程	equation of continuity
连续基础	continuous foundation
连续级配	continuous grading
连续加荷固结试验	continual loading test
连续加荷固结仪	continuously loaded oedometer
连续加热试验	continuous heat test

连续介质	continuous medium, continuum
连续介质力学	continuum mechanics
连续梁	continuous beam
连续墙导沟	leading trench of diaphragm wall
连续取岩芯	continuous coring
连续取样	continuous sampling
连续蠕变	continuous creep
连续施工	serial construction
连续速度测井	continuous velocity log
连续条件	continuity condition
连续性微分方程	differential equation of continuity
连续性原理	principal of continuity
连续桩成墙	continuous piled wall
联合基础	combined footing
梁式承台	pile cap as beams
梁式基础	beam foundation
量表	dial gauge, clock gauge
量测精度	accuracy of measurement
量测误差	measuring error
量测仪器	measuring device
量程	measuring range
量纲比	dimensional ratio
量纲分析	dimensional analysis
量力环	load ring, proving ring
量筒	dosimeter, graduated cylinder
料场勘探	borrow exploration
料堆	random bulk
料石	work stone
烈度测定	intensity determination
烈度调整	intensity adjustment
烈度分布	intensity distribution
烈度评定	intensity assessment
裂缝	crack

裂缝反射　crack reflection
裂缝灌浆　crack grouting
裂缝扩展速率　rate of crack propagation
裂片　splinter
裂纹扩展　crack growth
裂隙　fissure
裂隙糙度系数　joint roughness coefficient
裂隙水　crack water, crevice-water, fissure water
裂隙岩石　fissured rock
裂隙张开度　fissure aperture
邻桩　neighbouring pile
临界标准贯入击数　critical number of blows of standard penetration test
临界分析　threshold analysis
临界浮动梯度　critical floatation gradient
临界荷载　critical load(ing)
临界孔隙比　critical void ratio
临界粒度　critical particle size
临界密度　critical density
临界面　critical surface
临界频率　critical frequency
临界坡度　critical slope
临界水力梯度　critical hydraulic gradient
临界物态轨迹　critical state locus
临界物态模型　critical state model
临界物态能量理论　critical state energy theory
临界物态强度　critical state strength
临界物态土力学　critical state soil mechanics
临界逸出梯度　critical exit gradient
临界圆　critical circle
临界值　marginal value
临界桩长　critical length of pile
临界阻尼　critical damping
临空面　free face

临时槽支撑	temporary trench support
临时加固	temporary strengthening
临时岩体支护	temporary rock support
临时支撑	temporary support
临塑荷载	critical edge pressure
淋溶	eluviation
淋溶层	A-horizon
淋溶效应	leaching effect
淋余土	pedalfer
磷灰石	apatite
灵敏度	degree of sensitivity, sensibility, sensitivity
灵敏度分析	sensibility analysis
灵敏黏土	sensitive clay
零时水坡线	zero isochrone
零位指示器	null indicator
流变性	rheological behaviour (of mortar or slurry)
流变学	rheology
流变学模型	rheological model
流槽	flow channel, flow lane
流动崩坝	flowing avalanche
流动层	flow layer
流动法则	flow rule
流动破坏	flow failure
流动曲线	flow curve
流动土	yielding soil
流动性地基	flowing ground
流动指数	flow index
流函数	stream function
流径	flow path
流量	discharge
流量计	flow gauge
流沙层	mobile ground

流沙地基	running ground
流砂	drift sand, flowing sand, quick sand, running sand
流砂地层	fluidised bed
流水抽气器	aspirator
流水压力	pressure of water flow
流速	rate of flow
流体测沉计	hydraulic levelling device
流体重度	specific weight for fluid
流土	soil flow
流网	flow net
流纹岩	liparite, rhyolite
流纹岩的熔岩流	rhyolitic lava flow
流线	flow line
流岩	flow rock
流值	flow value
流状构造	flow structure
硫酸钡试验	barium sulphate test
硫酸盐矿渣水泥	slag-sulfate cement
隆起	upheave
隆起标桩	heave stake
隆起破坏	failure by heaving
隆胀	heave
隆胀力	heave force
漏斗	funnel
漏浆	leakage of mortar
漏水地基	leaky foundation
露头	outcrop
卤水	brine
卤水池	brine pond
卤水存储	brine storage
陆上沉积物	subaerial deposit
陆上沉井	land caisson
陆缘沉积	epicontinental sedimentation

滤布　filter cloth
滤层　filter layer, filtering layer
滤出物　leachates
滤井　well filter
滤流　filtering flow
滤水毡垫　filter mat
滤网式井点　screen well point
滤纸排水　filter paper drain
路边缘石　shoulder curb
路碴　road-metal
路堤　embankment
路堤沉陷　settlement of embankment
路基　road bed, roadbed, subgrade
路基断面　roadbed section
路肩线　shoulder line
路肩斜度　shoulder pitch
路面　pavement
路面抽水作用　pavement pumping
路堑　cutting
路堑边坡　cut slope
路线踏勘　route reconnaissance
吕荣单位　Lugeon unit
吕荣试验　Lugeon test
铝氧八面体　aluminum octahedron
履带式铲运拖拉机　cat and can
履带式风动钻机　air track drill
绿泥石　chlorite
卵石　cobble, gravel cobble, shingle, pebble
乱砌圬工　random masonry
乱切毛石　random rubble
乱石护坡　riprap
乱石墙　random rubble wall
伦敦黏土　London clay
轮胎压路机　pneumatic tyred roller

罗德参数　Lode's parameter
罗德角　Lode's angle
罗德数　Lode's number
螺钉　screw
螺钉锚固　screw anchor
螺丝端杆锚具　thread anchorage
螺旋板载荷试验　screw plate test
螺旋掏槽　spiral cut
螺旋压板载荷试验仪　field compressometer
螺旋桩　auger pile, screw pile
螺旋钻　auger drill, helical auger, twist auger, earth auger
螺旋钻探　auger boring
螺旋钻头　auger bit
洛杉矶磨损试验　Los Angeles abrasion test
洛氏硬度计　Rockwell apparatus
洛氏硬度试验机　Rockwell hardness tester
洛氏硬度值　Rockwell hardness number
洛氏针入硬度　Rockwell indentation hardness
洛阳铲　drive pipe, Loyang spoon, Luoyang spoon
落锤　drop hammer
落锤冲击贯入仪　drop-impact penetrometer
落锤试验　fall-cone test
落距　drop, height of drop, fall
落水洞　lime sink, dolinae, doline, aven
落水管　leader

M

马蓝黄土　Malan loess
马氏扁式松胀仪　Marchetti's flat dylatometer
埋藏阶地　buried terrace

埋入长度　embedded length
埋入式结构物　buried structure
埋深　depth of embedment
埋深比　depth ratio
埋置　embedment
埋置深度　laying depth, embedded depth
麦德维捷夫地震烈度表　Medvedev scale
麦卡利地震烈度表　Mercalli scale
麦克斯韦尔模型　Maxwell model
脉冲发生器　pulse generator
脉冲荷载　impulse load, pulse load
脉冲应力　pulse stress
曼代尔—克雷尔效应　Mandel-Cryer effect
慢剪试验　slow shear test, S-test
漫滩沼泽　backswamp
盲沟　blind drain, blind subdrain, french drain
盲沟排水　blind drainage
毛玻璃　frosted glass
毛料石　roughing stone
毛面岩石　rock block with rough and multiple
毛石　free stone, quarry stone, rubble
毛石挡土墙　rubble retaining wall
毛石工程　rubble work
毛石混凝土基础　stone-concrete footing
毛石基础　stone footing
毛石集料　rubble aggregate
毛石砌体　rubble stone masonry
毛石墙　rubble wall
毛石圬工　ordinary masonry
毛细带　zone of capillarity
毛细管间隙　capillary interstice
毛细管水　capillary water
毛细管水饱和　capillary saturation

毛细管水饱和带　zone of capillary saturation
毛细管水带　capillary zone
毛细管水上升边缘　capillary fringe
毛细管水上升高度　capillary height
毛细管水上升高度试验　capillary lift test
毛细管水势　capillary potential
毛细管水压力　capillary pressure
毛细管水移动　capillary migration
毛细管水张力　capillary tension
毛细管仪　capillarimeter
毛细管作用　capillarity
毛细裂缝　hair crack
锚板　anchor plate
锚定岸壁　anchorage bulkhead
锚定板墙　anchor slab wall
锚定板式挡土墙　tieback retaining wall
锚定板桩墙　anchored sheet pile(piling) wall
锚定土　anchored earth
锚定物　anchor
锚定装置　anchorage device
锚定作用　anchorage
锚杆　anchor bar, anchor rod, anchor tie, bolt, tieback
锚杆冲击机　rock bolter
锚杆挡墙　tieback wall anchored wall
锚杆应力　stress along bolt
锚固　bolting, anchoring
锚固长度　anchorage length
锚固点　anchorage point
锚固墩　anchored pier
锚固灌浆　anchor grouting
锚固浆液　anchor grout
锚固力　anchoring force
锚固应力损失　anchorage loss, loss due to an-

	chorage take-up
锚固粘结强度	anchorage bond strength
锚筋	beam pitman
锚孔	anchor hole
锚块	anchor block
锚拉式板桩墙	prestressed anchor pile-slab wall
锚塞	anchoring plug
锚栓	anchor bolt
锚索	anchor rope, anchorage cable
锚头	anchor head
锚桩	anchor pile
冒顶	cave in, fall of ground
梅花桩	quincuncial piles
梅纳旁压仪	Menard pressure meter
梅耶霍夫承载力公式	Meyerhof's formula
每分钟击数	blows per-min
每击贯入度	penetration per blow
美国公路局土的分类法 [BPRCS]	Bureau Public Road Clarification System
镁氧八面体	magnesium octahedron
镁质石灰岩	magnesian limestone
蒙蔽工程	shield engineering
蒙脱石	askanite, montmorillonite, smectite
糜棱岩	mylonite
米赛斯屈服条件	Mises yield condition
泌水	bleed
泌水路径	bleed path
密闭状态养护	sealed up concrete curing
密度	density
密度测定器	density gauge
密度计	densimeter
密度试验	density test
密度探测仪	density of probe

密度指数 density index
密尔(单位) mil
密封灌浆 seal grouting
密排桩 close pile
密砂 dense sand
密实度 compactness
密实装药 tight loading
幂法 power method
面板 facing
面积比 area ratio
面积仪 planimeter
明德林解答 Mindlin's solution
明矾黏土 alum clay
明沟 ditch, open drain
明挖 open excavation
明挖法 open cut method, out and corer method
明挖基础 open cut foundation
明挖隧道 open cut tunnel
模量比 modulus ratio
模量比效应 modular ratio effect
模拟级配 modeling gradation
模拟试验 simulated test
模型比例尺 model scale
模型律 model law
模型试验 model test
膜电位 membrane potential
膜式织物支撑 membrane type fabric support
摩擦角 friction angle
摩擦系数 coefficient of friction, friction coefficient
摩擦圆法 friction circle method
摩擦滞后 friction lag
摩擦桩 floating pile, friction pile, skin friction pile

摩擦桩基础　floating pile foundation
摩擦阻力系数　frictional resistance coefficient
摩擦阻尼　friction damping
摩尔包线，强度包线　Mohr's envelope
摩尔—库仑定律　Mohr-Coulomb law
摩尔—库仑理论　Mohr-Coulomb theory
摩尔—库仑模型　Mohr-Coulomb soil model
摩尔—库仑准则　Mohr-Coulomb criterion
摩尔强度包线　Mohr strength envelope
摩尔硬度计　Mohr scale of hardness
摩尔圆　Mohr's circle
摩尔圆包线　envelope of Mohr's circles
摩根斯坦法(边坡稳定分析)
　　Morgenstern method of slope stability
摩阻比　friction-resistance ratio
摩阻力　frictional resistance, drag
摩阻损失　friction loss
磨蚀　abrasion
磨圆度　psephicity
末端效应　end effect
母岩　mother rock, parent rock, matrix, country rock
母质层　C-horizon
木板桩　timber sheet pile, wooden sheet pile
木沉箱　timber pneumatic caisson
木铬胶　ligno-chrome gel
木桩　timber piles, picket
木桩碎裂　brooming

N

内部冲刷　internal scour
内部裂缝　internal crack

内含物　inclusion
内间隙比　inside clearance ratio
内径　internal diameter
内力　inner force, internal force
内力臂法　resisting moment arm method
内力重分布　internal force redistribution
内摩擦　internal friction
内摩擦角　angle of internal friction, internal friction angle
内摩擦系数　coefficient of internal friction
内黏聚力　internal cohesion
内排水　inner drainage
内时理论　endochronic theory
内围层　inlier
内压　internal pressure
内在渗透系数　intrinsic permeability coefficient
内支护　internal support
内阻尼　internal damping
钠膨润土泥浆　sodium bentonite mud
耐风化性　weatherability
耐交变应力性　resistance to alternate mechanical stress
耐扭力　torsibility
耐蚀性　corrosion resistance
耐水性　resistance to water
耐酸性　resistance to acid
难处理地基　difficult foundation
难处理土　troublesome soil
囊式密度计　balloon densimeter
囊式体积仪　balloon volumeter
囊式压力计　capsule-type pressure gauge
挠度　amount of deflection
挠度计　deflectometer
挠度曲线　deflection curve

挠曲应力　flexural stress
能量守恒定律　law of conservation of energy
尼龙　nylon
泥　mud
泥钉　soil nail
泥滑　mud wave
泥灰岩　marl
泥灰质黏土　marly clay
泥浆　pulp, slurry
泥浆泵　slurry pump
泥浆泵，污水泵　dredge pump, mud pump
泥浆材料　mud materials
泥浆槽　mud ditch
泥浆槽法　slurry trench method
泥浆层　mud layer
泥浆沉淀池　mud-setting pit
泥浆池　mud trap
泥浆处理　mud treatment
泥浆处理装置　mud treatment plant
泥浆反循环法　method of slurry reverse circulation
泥浆防渗墙　slurry cut-off, slurry trench wall
泥浆分析记录　mud analysis log
泥浆固壁　slurry stable wall
泥浆灌注　mud injection
泥浆护壁　slurry for preventing collapse of borehole, wall protection by slurry
泥浆护壁钻孔灌注桩　slurry bored pile
泥浆净化设备　slurry purification system
泥浆配方　mud formula
泥浆取样筒　slurry sampler
泥浆天平　mud balance
泥浆循环　mud circulation
泥浆正循环法　method of slurry direct circula-

tion

泥流 earth flow

泥流型滑坡 flow slide

泥盆纪 Devonian period

泥皮 mud cake

泥沙沉降 sediment deposition

泥沙粒径 diameter of sediment gain

泥沙筛孔粒径 sieve-aperture size for sediment

泥石流 debris flow, mud and rock flow, mud avalanche

泥炭 bog muck, peat, turf

泥炭土 moor soil

泥炭沼泽 peat bog

泥炭质土 peaty soil

泥土渗水能力 soil absorption capacity

泥岩 mudstone, siltstone

泥岩状结构 pelitic structure

泥质板岩 argillaceous slate

泥质沉积 argillaceous sediment

泥质砂岩 argillaceous sandstone

泥质岩 pelite

泥质页岩 argillaceous shale

拟合法 fitting method

拟塑性流体 pseudo-plastic fluid

逆断层 reverse fault

逆作法 reverse building method

年降雨量 annual precipitation

年径流量 annual runoff

黏弹特性 viscoelastic property

黏弹性 viscoelastic behavior, viscoelasticity

黏弹性地基 viscoelastic foundation

黏弹性模型 viscoelastic model

黏度计 cohesiometer

黏附 adhere

黏附性 adhesion character
黏滑断层 visco slip fault
黏聚力 cohesion
黏粒葱皮粘结 clay onion-skin bond
黏粒灌浆 clay grouting
黏粒含量 clay content
黏粒粒组 clay fraction
黏壤土 clayey loam
黏塑性流体 viscoplastic fluid
黏土 clay
黏土板岩 clay slate
黏土防渗层 earth membrane
黏土分级评价法 method of grading mud-making clay
黏土膏 clay puddle
黏土滑坡 clayslide
黏土混凝土 soil concrete
黏土夹层 clay parings
黏土胶结物 clay binder
黏土块 clay chunks
黏土矿物 clay mineral
黏土脉 clay vein
黏土片晶 clay platelet
黏土铺盖 clay blanket
黏土砂岩 clay sandstone
黏土水泥灌浆 clay-cement grouting
黏土透镜体 clay lens
黏土完全软化强度 fully softened strength of clay
黏土心墙 clay core
黏土岩 clay rock, claystone
黏土质页岩 clay shale
黏性泥沙 cohesive sediment
黏性土 clayey soil, cohesive soil

黏质粉土	clayey silt
黏质砂土	clayey sand
黏滞流	viscous flow
黏滞系数	coefficient of viscosity
黏滞性	tenacity, viscosity
黏滞阻尼	viscous damping
黏着系数	adhesion factor, coefficient of adhesion
黏着作用	adhesion
碾压	roller compaction
碾压遍数	number of roller passes
碾压法	roller compression, compaction by rolling
碾压机械	compacting machinery
碾压设备	compaction plant
碾压试验	rolling compaction test, rolling trial
碾压试验填土	test fill
碾压填土	rolled-earth fill
碾压土坝	rolled fill earth dam
捏	pinch
凝固剂	setting up agent
凝固强度	setting strength
凝固时间	setting time
凝灰岩	ashstone, tuff
凝灰岩水泥	tufa cement
凝胶	gel
凝结	setting
凝结膨胀	setting expansion
凝结水	condensation water
凝聚剂	coagulant
凝聚作用	agglomeration, coagulation
牛顿流体	Newtonian fluid
牛顿模型	Newton model
牛顿碰撞理论	Newton's theory of collision

牛顿液体 Newtonian liquid
牛轭湖 bayou(lake), oxbow lake
扭剪比 torsion-shear ratio
扭剪法 torsion-shear process
扭剪试验 torsional shear test
扭剪仪 torsion shear apparatus
扭矩 torque
扭矩系数 torque coefficient
扭力 torsional force
扭弯比 torsion-bending ratio
扭转变形 torsional deformation
扭转失稳 torsional buckling
扭转试验机 torsional tester
扭转应力 torsional shearing stress
扭转应力函数 stress function of torsion
扭转振动 torsional vibration
纽马克感应图 Newmark's(influence)chart
浓浆 thick grout
女儿墙 parapet wall
挪威黏土 Norwegian clay

O

欧拉断裂应力 Euler crippling stress
欧拉平衡微分方程 differential equilibrium equation of Euler
偶极矩 dipolar moment
偶极离子 dipolar ion
偶极子 dipole
耦合振动 coupled vibration

P

爬径长度　creep-path length
排除污染　decontamination
排泥泵　sludge pump
排气阀　air release valve, air bleed valve
排渗特性　drainage characteristics
排水　drainage
排水不良　impeded drainage
排水层　drainage layer
排水出口　drainage outlet
排水垫层　drainage blanket
排水法　drainage method
排水反复直剪试验　drained repeated direct shear test
排水沟　drain trench
排水固结法　consolidation by dewatering
排水管　drain
排水灌浆　displacement grouting
排水加荷　drainage loading
排水剪切试验　drained shear test
排水井　drainage sump, drainage well, well drain
排水孔　discharge orifice
排水滤层　drainage filter
排水路径　drainage path
排水三轴试验　drained triaxial test
排水砂井　sand drain
排水砂桩　drain pile
排水速度　discharge velocity
排水瓦管　drainage tile
排水系统　drainage system

排水下沉法　method of drained sinking
排水纸板　cardboard wick
排水钻孔　drilled drain hole
排污河流　receiving stream
排桩　piles in row, row of piles
盘底桩　disk pile
盘形滚刀回转切割岩石试验　rotary cutting test on the rock by the disc rolling cutter
庞塞莱特图解法　Poncelet graphical construction
旁侧导坑法　core-leaving method
旁压试验　lateral compression test, pressuremeter test(PMT)
旁压仪　pressuremeter
旁压仪极限压力　pressuremeter limit pressure
旁压仪模量　modulus of pressure meter, pressuremeter modulus
抛砂速度　ramming speed
抛石，堆石　enrockment
抛石工程　riprap work
抛石护岸　riprap protection
抛石护坡　riprap protection of slope
抛石基床　rubble-mound foundation
抛掷爆破　throwing blasting
炮眼　shot hole
炮眼间距　spacing between blast holes
配浆管路　grout distribution line
配筋率　reinforcement ratio
配料设备　batching equipment
配重下沉法　subsiding by matching weights
喷出岩　effusive rock, eruptive rock, extrusive rock
喷浆　gunite, spout of stock

喷浆机　shotcrete machine
喷锚法　shotcrete and rock bolt
喷锚支护　combined bolting and shotcrete
喷砂　sand-blasting
喷砂处理　sand blasting, treatment by sand-bath
喷砂处理法　sand-blasting method
喷射泵　jet pump
喷射杆　jet stem
喷射高度　jetting height
喷射管　jet mixer
喷射灌浆　jet grouting
喷射灌浆钻杆　jet grouting stem
喷射混凝土　air-blown concrete, pneumatically applied concrete, shotcrete, sprayed concrete
喷射混凝土外加剂　shotcrete admixture
喷射混凝土支护　shotcrete support
喷射井点　eductor well point, ejector well point
喷射井点系统　eductor well point system
喷射式钻孔机　jet boring machine
喷水冒砂　mud spouts
喷头　sprayer
喷嘴　jet nozzle
膨壳式锚杆　expansion shell bolt
膨润土　amargosite, bentonite
膨润土灌浆　bentonite grouting
膨润土浆液　bentonite slurry
膨润土泥浆　bentonite mud(grout)
膨润土水泥浆　bentonite cement grout
膨胀　expansion
膨胀度　degree of expansion
膨胀力　expansive force, swelling force

膨胀率　percentage of bed-expansion, rate of expansion, swelling ratio
膨胀率试验　swelling rate test
膨胀模量　modulus of dilatation
膨胀黏土　expanded clay
膨胀势　potential swell
膨胀试验　swelling test
膨胀土　dilatable soil, expansive soil, swelling soil
膨胀土地基加固　consolidation of expanding soil
膨胀系数　coefficient of swelling
膨胀性　expansibility
膨胀压力　expansive pressure, inflation pressure
膨胀页岩　expanded shale
膨胀仪　dilatometer, expansion apparatus
膨胀应变指数　swelling strain index
膨胀珍珠岩　expanded pearlite
劈理　cleavage
劈裂　spalling, split
劈裂灌浆　fracture grouting
劈裂抗拉强度　split tensile strength, splitting tensile strength
劈裂抗拉试验　split tensile test
劈裂强度　splitting strength
劈裂试验　Brazilian test, split test, splitting test
劈裂受拉强度　tensile splitting strength
疲劳　fatigue
疲劳断裂　fatigue fracture, repeated stress fatigue
疲劳荷载　fatigue loading
疲劳剪切仪　fatigue shear apparatus

疲劳界限	fatigue limit
疲劳破坏	fatigue failure
疲劳强度	fatigue strength
片间粘结	intersheet bonding
片晶	platelet
片麻石	gneiss
片岩	schist
片状成分	flaky constituents
片状节理	sheet jointing
片状颗粒	flake-shaped particle
片状砂岩	schistous sands
偏光显微镜	polarizing microscope
偏湿压实	wet compaction
偏位	deviation
偏位桩	deflected pile
偏心荷载	eccentric load, nonaxial load
偏心荷载基础	eccentrically loaded footing
偏心集中荷载	single non-central load
偏心距	eccentricity
偏心影响系数	influence factor for eccentricity
偏应力	deviator stress
偏应力状态	deviatoric state of stress
漂积层	erratic form
漂砾	erratic block, boulder
漂石	erratic
漂石黏土	boulder clay
贫黏土	lean clay
频率特性曲线	frequency-response curve
频率-振幅曲线	frequency-amplitude curve
频谱分析	frequency analysis
平板仪安装	setting of plane-table
平板载荷(荷载)试验	loading plate test, plate loading test
平扁石	penn stone

平差高程　adjusted elevation
平硐　side drift
平衡含水量　equilibrium moisture content
平衡锥　balance cone
平接　end joint
平截面假定　plane section assumption
平均比重(相对密度)　mean specific gravity
平均沉降　average settlement
平均海平面　mean sea level
平均厚度法　average thickness method
平均接触压力　average contact pressure
平均抗拉强度　average tensile strength
平均粒径　average (grain) diameter, mean diameter
平均模量　average modulus
平均应力　average stress
平均有效应力　mean effective stress
平均主应力　mean principal stress
平面变形　plane deformation
平面波　plane wave
平面滑动　plane slide
平面破坏　plane failure
平面应变　plane strain
平面应变拉伸试验　plane strain extension test
平面应变试验　plane strain test
平面应变问题　problem of plane strain
平面应变压缩试验　plane strain compression test
平面应变仪　plane strain apparatus
平面应变状态　plane strain state
平面应力　plane stress
平面应力问题　problem of plane stress
平台桩　platform pile
平土机　grader
平巷地压　rock pressure in horizontal roadway

平行断面法	parallel section method
平行级配法	parallel grading method
平移断层	displacement fault
瓶颈效应	ink bottle effect
坡底圆	base circle
坡顶裂缝开展深度	tension crack depth on the top of slope
坡度比	ratio of slope
坡度变点	point of change of gradient
坡度突变	knuckle
坡积物	cliff debris, slope wash
坡角	angle of slope
坡脚	base of slope, slope toe, toe of slope
坡脚开挖	toe excavation
坡脚墙	slope toe wall
坡脚圆	toe circle
坡口角度	angle of bevel
坡面冲蚀	slope surface erosion
破坏包线	envelope of failure, failure envelope
破坏轨迹	failure locus
破坏荷载	breaking load, failure load, load at rupture
破坏机理	failure mechanism
破坏假说	failure hypothesis
破坏面	failure plane, failure surface
破坏模式	mode of failure
破坏强度	failure strength
破坏区	failure zone
破坏试验	failure test
破坏条件	failure condition
破坏效应	shattering effect
破坏应变	failure strain
破坏应力	failure stress
破坏圆	circle of failure

破坏指数　failure index
破坏状态　collapse state
破坏准则　failure criterion
破裂角　angle of rupture
破裂强度　rupture strength
破裂应力　rupture stress
破碎　spall
破碎带　fracture zone
破碎料　crushed-run aggregate
破碎压力　breakdown pressure
剖面　section
剖面图　profile
剖视图　sectional view
铺盖　blanket
铺盖灌浆　blanket grouting
铺土厚度　lift thickness
铺网法　fabric sheet reinforced earth
普拉格模型　Prager model
普朗特承载力理论　Prandtl bearing capacity theory
普朗特塑性平衡理论　Prandtl plastic equilibrium theory
普罗克特击实曲线　Proctor compaction curve
普罗克特击实试验　Proctor compaction test
普罗克特针测含水量试验　Proctor needle moisture test
普罗托吉雅可诺夫数　Protodyakonoves' number
普氏系数（岩石强度系数）　Protodyakonov number
普氏压力拱理论　Protodyakonov's theory

Q

齐发爆破　simultaneous blasting
起爆　detonation
起吊应力　handling stress
起伏地形　rolling topography
起重机安全装置　safety devices of crane
气锤　pneumatic hammer
气举　air lift
气举泵　airlift pump
气幕法　air-curtain method
气枪　air gun
气水结合面　air-water surface
气体比重计　aerometer
气隙比　air space ratio, air void ratio
气相　gaseous phase
气压沉箱　compressed-air caisson, pneumatic caisson
气压盾构法　shielding with air pressure
气压法　compressed-air method
气压式沉降仪　pneumatic settlement cell
气压式孔隙水压力仪　pneumatic piezometer
气压桩　pneumatic pile
气闸　air-lock
弃方　spoil
弃土　waste
弃土堆　banquette
弃土区　disposal area
汽车钻机　wagon drill
汽锤打桩机　ram steam pile driver
砌体挡墙　masonry retaining wall
砌体工程　masonry

千斤顶　jack
千斤顶标定　jack calibration
千斤顶举升力　jacking force
千斤顶力损失　jack losses
千枚岩　phyllite
钎探　rod sounding
铅垂线　plumbline
前冰期沉积物　preglacial deposit
前护墙　apron wall
前屈曲平衡状态　pre-buckling equilibrium state
前震　foreshock
前震旦纪　Pre-Sinian period
潜孔锤　down-hole hammer
潜水　subsoil water, subsurface water
潜水井　latent-water well
潜水面　water plane, phreatic surface
潜水水位　phreatic water level(surface)
潜堰　barrier
潜在不稳定性　latent instability
潜在滑动面　potential surface of sliding
潜在破坏面　potential failure surface
潜在渗入强度　potential infiltration rate
浅层处理　shallow treatment
浅层地基加固法　stabilization of superficial subgrade
浅层地下水　shallow ground water
浅层灌浆　shallow grouting
浅层滑坡　shallow failure
浅层渗透　shallow percolation
浅层土加固　surface soil stabilization
浅层压实　shallow compaction
浅层钻孔沉桩　shallow sinking piles with bores
浅基础　shallow foundation
浅基防护　protection of shallow foundation

浅井　shallow shaft
浅井注水　shallow well injection
浅孔爆破　shallow hole blasting
浅孔液压爆破　shallow hole hydraulic blasting
浅埋基础　shallow footing
浅埋隧道　shallow tunnel, shallow-buried tunnel
浅源地震　shallow-focus earthquake
欠固结　underconsolidation
欠固结黏土　underconsolidation clay
欠固结土　underconsolidated soil
欠挖　shallow cut, shallow dredging
堑壕法　trench method
嵌固长度　length of restraint
嵌岩管柱轴向承载力　axial bearing capacity of drill caisson embedded in bedrock
嵌岩柱　anchored in rock piles
饿台　berm
强度　strength
强度包线　strength envelope
强度变形特性　strength-deformation characteristic
强度储备系数　strength safety coefficient
强度等级　strength grade
强度分级　strength class
强度分析　strength analysis
强度各向异性指标　strength anisotropy index
强度计算　strength calculation
强度降低系数　strength reduction factor
强度理论　strength theory, theory of strength
强度-龄期关系　strength-aging relationship
强度-密度比　strength to density ratio, strength-density ratio
强度模型　strength model

强度曲线　intensity curve
强度试验　strength test
强度特征值　strength characteristic
强度条件　condition of strength, strength condition
强度退化　retrogression of strength
强度指标　index of strength
强度终值　ultimate strength value
强风化黏磐土　nitosol
强夯法　dynamic consolidation
强化阶段　strengthened stage
强黏性土　tenacious clay
强迫下沉　enforced settlement
强迫振动　forced vibration
强迫振动法　method of forced vibration
强震　macroseism
强震带　pleistoseismic zone
墙背排水设施　back drain
墙基　wall foundation
墙摩擦角　angle of wall friction
墙摩擦力　wall friction
墙黏着力　wall adhesion
墙下筏形基础　raft foundation under walls
墙下条形基础　strip foundation under walls
乔斯登硅化加固　Joosten process
桥墩　bridge pier, pier
桥台　abutment
切层滑坡　insequent landslide
切割射流　cutting jet
切土　soil cutting
切土环刀　circular soil cutter
切土筒　clay cutter
切线刚度　tangent stiffness
切线模量　tangent modulus

切线模量理论　tangent modulus theory
切线摩擦力　tangent friction force
切向力　tangential force
切应变　tangential strain
切应力　tangential stress
亲水胶体　hydrophilic colloid
亲水性　affinity of water
侵入岩　intrusive rock, irruptive rock
侵蚀　penetration of slag
侵蚀面　erosion surface
侵蚀性水　corrosive water, aggressive water
氢氧离子　hydrated ion
轻便触探试验锤击数 N_{10}　light sounding test blow count
轻便剪切仪　portable shear apparatus
轻石土　pumice soil
轻质土　light soils
倾倒破坏　toppling failure
倾覆力　overturning force
倾覆力矩　overturning moment, tilting moment, upsetting moment
倾覆破坏　tilting failure
倾覆稳定系数　stability factor against overturning
倾覆稳定性　overturning stability, tilt stability
倾角　angle of inclination
倾角　dip angle, dip, angle of obliquity, amount of inclination
倾角计算法　inclination method
倾角量测仪　tiltmeter
倾向滑断层　dip slip fault
倾向节理　dip joint
倾斜度　obliquity
倾斜断层　inclined fault

倾斜观测	inclination observation
倾斜荷载	inclined load, tilt load
倾斜基础	sloping foundation
倾斜监测装置	tilt monitoring device
倾斜岩层	tilted stratum
倾斜仪	dipmeter
清华弹塑性模型	Tsinghua elastoplastic model
清孔	borehole cleaning
清孔机具	clean-out tools
清孔钻	clean-out auger
清理场地	grubbing
清水护壁	wall stabilization with clear water
穹隆式基础	dome foundation
球根桩	bulb pile
球窝接头	ball and socket joint
球应力	spherical stress
球状颗粒	spherical particles
区域工程地质	regional engineering geology
区域水文地质学	regional hydrogeology
区域水文学	regional hydrology
区域性沉降	regional settlement
区域性地下水位下降	descent of regional ground water level
区域性土	zonal soil
曲度	sharpness
曲率系数	coefficient of curvature
曲线拟合	curve fitting
曲线桩	curve stake
屈服	yield
屈服点	point of yielding, yield point
屈服轨迹	yield locus
屈服函数	yield function
屈服极限	limit of yielding
屈服阶段	plastic stage

屈服面　yield surface
屈服强度　yield strength
屈服应力　yield stress
屈服应力模型　yield stress model
屈服值　yield value
屈服准则　yield criteria
屈曲理论　theory of buckling
渠道　channel
渠道化　canalization
取水样器　water sampler
取土　borrow
取土管　soil sampling tube
取土坑　borrow pit
取土器　soil sampler
取土筒　soil sample barrel
取芯钻头　rock core bit
取样　sampling action, soil sampling
取样长度　sampling length
取样法　method of sampling
取样观测　sampling observation
取样管　sampling tube
取样盒　soil sample box
取样井　sampling well
取样扰动　sampling disturbance
取原状土样　undisturbed soil sampling
圈梁　periphery beam
全断面法　full face method
全孔隙压力比　full pore-pressure ratio
全量分析　gross analysis
全息干涉测量　holographic interfrometry
全新世　holocene epoch
全压式盾构　shield with compressed air
缺陷桩　imperfect pile
群桩　group piles, multiple pile, pile group

群桩沉降比　settlement ratio of stability
群桩沉降量　settlement of pile group
群桩抗力　resistance of piles
群桩排架　cluster bent
群桩深度效应　depth effect of pile group
群桩竖向极限承载力　vertical ultimate capacity of pile groups
群桩相互作用　pile group interaction
群桩效率　group efficiency
群桩效应　pile group action of pile
群桩折减系数　reduction factor for pile in group

R

染色法　dyeing
壤土(垆姆，亚黏土)　loam
扰动比　disturbance ratio
扰动度　disturbance degree
扰动土样　disturbed soil sample
扰动指数　disturbance index
绕坝渗漏　seepage around abutment
绕渗　by-pass seepage
热变形　thermal deformation
热处理　heat treatment
热传导　heat transfer
热对流　heat convection
热辐射　heat radiation
热固结　thermal consolidation
热敏的　sensitive to heat
热摩奇金法　Zemochkin's method
热黏弹性理论　theory of thermoviscoelasticity
热容量　heat capacity

热渗系数	coefficient of thermoosmotic transmission
热生应变	thermally induced strain
热粘合	heat bonding
人工边坡	man-made slope
人工地基	artificial foundation(ground)
人工防冲铺盖	armor
人工加固土	artificially improved soil
人工开挖	hand-dug
人工取样	manual sampling
人工填土	artificial fill(soil)
人工挖孔灌注桩	pile with man-excavated shaft
人力冲击钻探	percussion hand boring
人力夯	hand rammer
人力挖井	hand-dug well
人力钻探	manual boring
人为侵蚀	human erosion
人为震害	man-made seismic hazards
人造纤维	man-made fiber
任意假定	arbitrary assumption
任意力系的简化	reduction of an arbitrary force system
容水量	water bearing capacity
容许残留冻土层厚度	allowable thickness of residual frost layer
容许长细比	allowable slenderness ratio
容许沉降(量)	allowable settlement, permissible settlement, tolerable settlement
容许承载力	allowable bearing capacity, allowable bearing pressure, allowable bearing value
容许单桩荷载	allowable pile bearing load
容许荷载	admissible load, permissible load, safe load

容许极限　acceptable limit
容许黏着应力　allowable bond stress
容许偏斜　allowable deflection
容许使用荷载　allowable working load
容许土压力　allowable soil pressure, permissible soil pressure
容许位移　allowable displacement
容许误差　allowable error
容许相对变形　allowable relative deformation
容许压力　allowable pressure
容许应变　admissible strain
容许应力　admissible stress, allowable stress, permissible stress
容许应力法　permissible stress method
容许应力设计法　permissible stress design method
容许振动加速度　allowable vibration acceleration
容许振幅　allowable amplitude
溶洞　solution cavity
溶洞泉　solution channel spring
溶洞岩石　cavern rock
溶度计　lysimeter
溶沟　solution groove
溶解　dissolution
溶解度　solubility
溶解度曲线　solubility curve
溶解盐　dissolved salts
溶滤　lixiviation
溶滤变形系数　coefficient of deformation due to leaching
溶蚀槽　solution channel
溶蚀洞穴　solution cave
溶蚀盆地　dissolution basin

溶液灌浆	solution grout, solution injection
溶液浓度	solution concentration
溶质偏析	soluble segregation
溶质吸力	solution suction
熔岩	lava
熔岩灰	lava ash
熔岩流	lava flow
熔岩通道	lava tube
融沉土	sagging soil
融沉系数	coefficient of thaw-subsidence
融沉因数	thaw subsidence factor
融化深度	thaw depth
融土	thawed soil
融陷	thaw collapse
融陷性	thaw collapsibility
柔韧性	pliability
柔性薄膜衬垫	flexible membrane liner
柔性底层	soft ground floor
柔性管道	flexible conduit
柔性荷载	flexible load
柔性基础	flexible foundation
柔性路面	flexible pavement
柔性墙	flexible wall
肉眼认别法	megascopic method
蠕变	time-yield, creep
蠕变沉降	creep settlement
蠕变函数	creep function
蠕变荷载	creep load
蠕变极限	creep limit
蠕变率	rate of creep
蠕变破坏	creep rupture
蠕变强度	creep strength
蠕变试验	creep test
蠕变速率	creep rate

蠕变损伤　creep damage
蠕变损伤变形　creep deformation
蠕变系数　coefficient of creep
蠕变性状试验　creep behaviour test
蠕变应变　creep strain
蠕变运动　creep motion
蠕动坡积物　creeping waste
蠕动线　line of creep
蠕滑断层　creeping rock mass
入射波　incident wave
入射角　angle of incidence
入渗点　infiltration point
软带　weak zone
软化　softening
软化剂　softening agent
软化系数　softening index
软基加固　consolidation of soft subsoil
软基加筋　reinforcement of soft foundation
软黏土　mild clay, myckle, soft clay
软弱地基　subgrade, poor subsoil, weak foundation, weak ground
软弱夹层　soft rock strata, weak intercalated layer
软弱结构面　weak structural plane
软弱面　plane of weakness
软土　soft soil, weak soil
软土的侧向挤出　lateral squeezing-out of soft soil
软土地基　soft clay ground, soft foundation, soft soil foundation
软岩　soft rock, weak rock
软岩层　soft formation
瑞典圆弧法　Swedish circle method
瑞利波　Rayleigh wave

弱风化层　weakly weathered layer
弱风化带　weakly weathered zone
弱胶结岩层　incompetent rock
弱结合水　loosely bound water, pellicular water
弱透水层　aquitard

S

三叠纪　Triassic
三合土垫层　bedding course
三合土基础　foundation made of materials
三合土墙　lime earth concrete wall
三角掏槽　triangle cut
三角形法　triangular method
三角洲　delta
三角洲交错层　deltic cross bedding
三角坐标图　ferett triangle
三维应力　three dimensional stress
三相图　three phase diagram
三相土　tri-phase soil
三向变形条件下的固结沉降　three-dimensional consolidation settlement
三重管旋喷法　triple-pipe chemical churning process
三轴剪切试验　triaxial shear test
三轴拉伸试验　triaxial extension test
三轴试验　triaxial test
三轴收缩试验　triaxial shrinkage test
三轴压力室　triaxial cell
三轴压缩试验　triaxial compression test
三轴仪　triaxial apparatus
三轴应力状态　triaxial state of soil

散体地压	earth pressure of loose ground
散体力学	mechanics of granular media
色卢铁解答	cerruti's solution
沙井	sand-filled drainage well
沙洲砾石	bar gravel
砂崩	sand avalanche
砂泵	sand pump
砂袋	sand bag
砂袋护坡	sandbag revetment
砂袋筑墙	sandbag walling
砂的相对密实度试验	sand relative density test
砂垫层	sand bedding course, sand blanket, sand cushion, sand mat, sandpad
砂垫层加固法	sand cushion stabilization method
砂堆比拟	sand heap analogy
砂沸	sand boiling, boiling
砂固结锚固	sand consolidation anchorage
砂灰比	sand cement ratio
砂浆	mortar, sand grouting, sand mortar
砂浆的应力-应变关系	stress-strain relationship of mortar
砂浆分层度	segregation of mortar
砂浆片剪切法	the method of mortar flake
砂浆强度等级	strength grading of mortar
砂井排水法	sand drain method
砂井真空排水法	sand drain vacuum method
砂类土	sandy soil
砂砾	sand and gravel
砂砾垫层	gravel-sand cushion
砂砾覆盖层	sand and gravel overlay
砂砾石	sandy gravel
砂砾石垫层	sandy gravel layer
砂粒分级	sand grading
砂粒含量	sand content

砂滤池　sand filter
砂滤法　sand filtration
砂率　percentage of sand
砂丘　dune
砂石比　sand-coarse aggregate ratio
砂石率　sand coarse aggregate ratio(S/A)
砂石桩　sand-gravel pile
砂土　sand, sand soil
砂土地基　sandy soil foundation
砂土液化　liquefaction of sand
砂屑岩　arenite, arenyte
砂型水泥　sand casting cement
砂岩　sandstone, arenaceous sediment
砂质垂直排水　sand chimney
砂质粉土　sandy silt
砂质海床　sandy seabed
砂质集料　sandy aggregate
砂质黏土　sandy clay, dauk
砂质石灰岩　sandy limestone
砂质页岩　sandy shale
砂桩　sand column, sand pile
砂桩船　sand piling barge
砂桩挤密　sand compaction pile
砂桩加固法　sand pile stabilization method
筛铲　sieve shovel
筛分　screen classification
筛分法　screening, sieve method
筛分分级　screen sizing, sieve classification, sieve sizing
筛分级配　screen size gradation
筛分检查　screening inspection
筛分粒度　screen size, sieve size
筛分粒度级　sieve fraction
筛分曲线　sieve curve

筛分试验　screen test, sieve test
筛分析　mesh analysis
筛号　size of mesh
筛孔　sieve mesh, mesh
筛孔尺寸　size of screen mesh
筛砂　sand screen
筛网　sieve net
筛析　sieve analysis
筛选　screen
筛余　material retained
筛余百分率　retained percentage
筛余物　retained material on the sieve, screen tailings, sieve residue
山地泥炭　mountain peat
山洪　freshet
山麓　piedmont
山麓冲积平原　bahada
山体压力　rock pressure in hill
山岩压力　rock pressure
闪长岩　diorite
扇形分布　sector distribution
扇形惯性矩　sectorial moment of inertia
扇形几何性质　sectorial geometric property
扇形剪应力　sectorial shear stress
扇形静面矩　sectorial statical moment of area
扇形理论　sector theory
扇形正应力　sectorial normal stress
扇形坐标　sector coordinate
扇形坐标系　sector coordinate system
上部结构和基础的共同作用分析
　　analysis on interaction of superstructure and foundation
上部结构—基础—地基共同作用分析
　　structure-foundation-soil interaction analysis

上层滞水	perched water, vadose water
上层滞水含水层	perched aquifer
上覆地层	overlying strata
上覆土	superimposed soil
上覆岩石	overlaying rock
上孔法	up-hole method
上坡	uphill
上游围堰	upstream cofferdam
烧失量	ignition loss
勺钻	bucket auger
蛇纹石	serpentine boulder
设备	device
设防烈度	design intensity
设计荷载	design load
设计压实层厚	design lift
设计准则	design criteria
射流泵	efflux pump, injection pump
射水沉桩	jetted pile
射水成桩	jet pile
射水冲桩	pile jetting
射水打桩法	jetting piling
射水打桩机	water jet driver
摄影测斜仪	photographic inclinometer
伸长比	elongation ratio
伸长计	extensometer
伸缩缝	expansion joint, shrinkage and tension joint
伸缩式扩孔机	telescopic bucket reamer
深泵井	deep pumped well
深部破裂面	underground rupture plane
深层沉降	deep settlement
深层沉降仪	deep settlement gauge
深层触探试验	deep sounding test
深层地基加固	stabilization of deep subgrade

深层灌浆	deep-seated grouting
深层滑坡	deep slide
深层混合搅拌桩机器	deep jet mixing pile machine
深层加固	deep consolidation
深层加密	deep densification
深层搅拌法	deep mixing method
深层搅拌桩复合地基	cement deep mixing composite foundation
深层蠕动	depth creep
深层渗透	deep percolation
深层石灰搅拌法	deep-lime-mixing method
深层土加固	deep soil stabilization
深层压实	compaction of deep bed, deep compaction
深成岩	abyssal rock, hypogene rock
深成岩体	pluton
深度效应	effect of depth
深度因素	depth factor
深海	abyssal sea
深海沉积	abysmal deposit, pelagic deposit
深海沉积物	abyssal sediment, deep sea deposit, deep-sea sediment
深海红土	abyssal red earth
深海黏土	abysmal clay
深海区	abysmal area, abysmal area region
深海软泥	abyssal ooze, deep ocean ooze
深海相	abyssal facies
深基础	deep foundation
深井泵	deep-well pump
深井法	deep well method
深井滤水管	screen of deep well
深开挖	deep cut, deep excavation
深孔爆破	long hole blasting, deep blasting

深孔爆破法	long-hole method
深孔灌浆	long hole grouting
深孔钻进	long hole drilling
深埋锚定桩	deadman
深源地震	anatectic earthquake
深钻孔,深层钻探	deep boring
沈珠江三重屈服面模型	Shen Zhujiang three yield surface method
渗出	seepage-off
渗径	leakage path
渗径	seepage path
渗坑	percolation pit
渗流	seepage
渗流控制	seepage control
渗流连续方程	continuity equation of seepage
渗流量	quantity of percolation
渗流量	seepage discharge
渗流量	seepage quantity
渗流模型	seepage-flow model
渗流区	vadose zone
渗流速度	seepage velocity
渗漏	influent seepage, oozing, percolation
渗漏井	leaking well
渗漏量	amount of leakage
渗漏水	leakage of water
渗漏损失	leakage loss
渗漏系数	leakage factor
渗滤	diffusion
渗滤试验	percolation test
渗溶作用	leaching
渗入量	infiltration capacity
渗入水	water of infiltration, water of percolation
渗入性灌浆	permeable grouting, seep-in grouting

渗水	penetration of water, water creep
渗水测试	water test
渗水井	filter-well, infiltration well, leaching well, soakage pit
渗水试验	infiltration test
渗透	infiltration, osmosis
渗透变形	seepage deformation
渗透变形试验	filtration erosion test
渗透灌浆	permeation grouting
渗透距离	penetration distance
渗透力	seepage force
渗透路径	infiltration path, path of percolation
渗透率	infiltration rate
渗透破坏	seepage failure
渗透区	permeable zone, pervious zone
渗透势	osmotic potential
渗透试验	permeability test
渗透水	osmotic water
渗透速度	infiltration velocity, velocity of permeability
渗透速率	percolation rate, rate of percolation
渗透损失	seepage loss
渗透吸力	osmotic suction
渗透系数	coefficient of infiltration, coefficient of permeability, percolation coefficient, permeability coefficient, seepage coefficient
渗透现象	osmotic phenomenon
渗透性	permeability, perviousness
渗透性能	penetration characterisitics
渗透压法	osmotic pressure method
渗透压力	osmotic pressure, seepage pressure
渗透仪	percolation apparatus, permeameter
渗透运动	osmose movement

生石灰　quicklime, unslacked lime
生石灰桩　unslacked lime pile
生物沉积　biogenic deposit
生物风化　biological weathering
生物化学岩　biochemical rock
生物界　organic sphere
生物软泥　bioslime
生物岩　biolith
声波打桩机　sonic pile driver
声波法探测　sonic prospecting
声发射测压计　acoustic piezometer
声发射监测　acoustic emission monitoring
声发射应变计　acoustic strain gauge
声法测井　acoustic(al) log
声幅测井　amplitude log
声频仪　audio equipment
声速测井　acoustic(al) velocity logging
声学法　acoustic method
声学探测　acoustic prospecting
绳索悬挂瓣式抓斗　cable-hung clamshell bucket
圣维南原理　Saint-Venant's principle
失稳极限荷载　limit-load of instability
失效石灰　dead lime
施工放样　setting out
施工荷载　site load
施工勘察　investigation during construction
施工现场总平面图　overall site layout
湿捣法　moist rodding
湿度　moisture
湿度计　hygrometer, psychrometer
湿度指数　wetness index
湿法粒径分析　wet mechanical analysis
湿法筛分　wet screening
湿分析法　wet analysis

湿化　slaking
湿化试验　slaking test
湿击法　moist tamping
湿密度　wet density
湿膨胀　moisture expansion
湿容重（表观密度）　wet unit weight
湿式喷射法　wet-mix spraying process
湿式凿岩机　water-fed rock drilling
湿试样　wet sample
湿土　wet material
湿陷量　collapsible settlement
湿陷起始压力　initial collapse pressure
湿陷系数　coefficient of collapsibility
湿陷性　collapsibility
湿陷性地基　collapsible subsoil
湿陷性黄土　collapsible loess
湿陷性黄土地基加固　consolidation of collapsing soil
湿陷性土　collapsible soil, water sensitive soil
湿胀　wet swelling
湿胀率　percentage bulking
十字板贯入仪　vane penetrometer
十字板剪力仪　vane shear apparatus
十字板剪切试验　vane shear test
十字板抗剪强度　vane strength
十字板仪　four-bladed vane, vane borer
十字臂沉降管　cross-arm settlement gauge
十字交叉条形基础　crossed strip foundation
十字形黏土冲切器　cruciform clay cutter
十字形钻　cross mouthed drill
十字钻头　cross bit
石钉　rock dowel
石膏　gypsum
石膏矿渣水泥　slag gypsum cement

石化作用　petrification
石灰活化性　lime reactivity
石灰加固　lime stabilization
石灰砂浆　lime mortar
石灰土　lime soil
石灰稳定土　lime stabilized soil
石灰系深层搅拌法　lime deep mixing method
石灰岩　limestone
石灰岩溶洞　limestone cave
石灰质土　calcareous soil
石灰柱　lime column, lime pile, quicklime pile
石灰桩法　lime pile method
石灰桩挤密　lime compaction pile
石蜡　paraffin
石料　stone material
石锚　rock anchor
石棉　asbestos
石器时代　anthroplithic age
石炭纪　Carboniferous period
石屑　gallet
石英　quartz
石英砂　quartz sand
石英砂岩　silicarenite
石英岩　quartzite
石质扇形地　rock fan
石质土　lithosol, stony soil
时间常数　time constant
时间对数拟合法　logarithm of time fitting method
时间分析　time analysis
时间固结曲线　time consolidation curve
时间效应　time effect
时间因子　time factor, T_V
时效　aging effect

实测现场强度　site strength
实测值　active measured value
实地试验　site trial
实际尺寸　full size
实际工程量　actual quantity
实际流体　real fluid
实际应用　practical applications
实例研究　case study
实体测定法　body measurement method
实体筏基　solid raft
实体式基础　block foundation
实在应力　actual stress
拾振器　geophone, oscillation pickup
蚀变　alteration
蚀变岩石　altered rock
矢径　radius vector
使用荷载　working load
使用年限　service life
示波器　oscillograph
示波仪　oscillometer
势函数　potential function
势降　potential drop
势能　potential energy
势能驻值原理　principal of stationary potential energy
试件　specimen, test piece, test specimen
试井　test shaft
试坑　trial pit
试坑渗透试验　pit permeability test
试验　experimentation
试验参数　experimental parameters
试验成果　test result
试验打桩　trial piling, trial driving
试验分散性　sample dispersion

试验荷载　testing load
试验加载设计　test loading design
试验设备　testing equipment
试验室土样拌合器　laboratory soil mixer
试验数据　experimental data
试验台　experimental table
试验填方　test embankment
试验土工学　experimental soil engineering
试验研究　experimental study
试桩　test pile, trial pile
试钻　trial boring
室内十字板试验　laboratory vane test
室内试验　laboratory test
室内土工试验　laboratory soil tests
室压　cell pressure
适应环境　acclimation
释放节理　release joint
收控爆破技术　controlled blasting technique
收缩比　shrinkage ratio
收缩沉降　shrinkage settlement
收缩度　degree of shrinkage
收缩缝灌浆　contraction-joint grouting
收缩集料　shrinkable aggregate
收缩开裂　shrinkage-induced cracking
收缩裂缝　contraction fissure, shrinkage crack
收缩模量　shrinkage modulus
收缩曲率　shrinkage curvature
收缩曲线　shrinkage curve
收缩试验　shrinkage test
收缩系数　coefficient of shrinkage, shrinkage coefficient
收缩性土　contractive soil
收缩-徐变关系　shrinkage-creep relationship
收缩应变　shrinkage strain

收缩应力	retraction stress, shrinkage stress
手动螺旋钻	hand-operated auger
手工钻探	hand boring
手提击实仪	portable compacter
手钻	hand auger
首震	preliminary shock
受剪面层作用	shear diaphragm action
受拉承载力	tensile capacity
受拉荷载	tensile load
受拉裂缝	tensile crack
受拉破坏	tensile failure
受拉区	tensile area, tensile region
受拉应力-应变曲线	stress-strain curve in tension
受污染土	contaminated soil
受压承载力	compression capacity
书堆组构	bookhouse fabric
疏干	drainage by desiccation, unwatering
疏干系数	depletion coefficient
疏浚标高	dredge level
疏密波	dilatational wave
输沙量	sediment load
熟石灰	hydrated lime
树根桩	root pile
树脂锚杆	resin bolting
竖管式测压计	standpipe piezometer
竖井	rising shaft, shaft well
竖井井口	shaft mouth
竖井身	shaft
竖向固结系数	coefficient of vertical consolidation
竖向截水体	chimney drain
竖向开挖法	vertical excavation method
竖向收缩率	vertical shrinkage

竖胀潜量　potential vertical rise(PVR)
竖直桩　vertical pile
竖桩　soldier pile
数量级　order of magnitude
数学模型　mathematical model
数值分析　numerical analysis
数值积分　numerical integration
数值岩土力学　numerical geomechanics
衰化参数　degradation parameter
衰化指数　degradation index
衰减　attenuation
衰减时间　decay time
衰减系数　coefficient of attenuation
衰减周期　decay period
栓塞灌浆(法)　packer grouting
双壁板桩围堰　double-well sheet pile cofferdam
双壁钢沉井　steel open caisson with double-shell
双壁钢丝网水泥沉井　open caisson with two shells of wire-mesh cement
双参数地基模型　two-parameter foundation model
双层地基　double-layer soil foundation
双层取样器　double tube sampler
双层岩芯管　double core barrel
双电层　double layer, electric double layer
双动汽锤　double-acting steam hammer
双对数坐标图　loglog plot
双峰形粒径曲线　bimodal curve
双价离子　divalent ion
双剪应力屈服模型　twin shear stress yield criterion

双面剪切试验	double shear test
双膜式土压力盒	double diaphragm pressure cell, double diaphragm pressure gauge
双排板桩围堰	double wall cofferdam, two-wall sheet-piling cofferdam
双桥式触探仪	double bridge type penetrometer
双曲线模型	hyperbolic model
双塞灌浆	double-packer grouting
双弯曲	tangent bend
双线性模型	bilinear model
双样固结试验	double-specimen oedometer test
双液法	two-fluid process, two-shot method
双振幅	double amplitude
双轴抗拉试验	biaxial tensile test
双轴应力状态	biaxial state of stress
水玻璃	sodium silicate, water glass
水玻璃溶液	water glass solution
水成沉积	hydatogen sediment
水成土	hydromophic soil, aqueous soil
水成岩	hydatogenous rock
水冲法	jetting process
水冲抛石	pierre perdue
水冲式贯入仪	wash point penetrometer
水冲刷	wash out
水冲钻孔	hydrauger hole
水冲钻探	water flush boring
水处理	water treatment
水的进入	ingress of water
水的密度	density of water
水底隧道	underwater tunnel
水分迁移	moisture migration
水分指数	moisture index
水封	water sealing

水工隧道 water tunnel, hydraulic tunnel
水合作用 hydration
水化矿物 hydrated mineral
水化膨润土 hydrated bentonite
水化热 hydration heat
水化学 aquatic chemistry
水灰比 water cement ratio
水解性 slaking property
水库岸坡稳定性 stability of reservoir slope
水力沉桩 sinking pile by water jet
水力冲击钻探法 hydraulic percussion method
水力冲填 hydraulic fill
水力劈裂 hydraulic fracture
水力劈裂法 hydraulic fracturing technique
水力坡度 hydraulic slope
水力梯度 hydraulic gradient
水力透水性 hydraulic permeability
水力压力盒 hydraulic load cell
水力钻探 jetting drilling
水利工程 water conservancy project
水流出逸口 outcrop of water
水流路径 water course
水铝英石 allophane, allophanite
水镁石 brucite
水膜理论 water film theory
水泥灌浆 cement grouting, cement injection
水泥加固 cement stabilization
水泥加固土 cement-stabilized soil
水泥浆 water cement slurry
水泥结石 harden grout film
水泥砂浆 cement mortar
水泥土 soil-cement material
水泥土加固 cement soil stabilization
水泥土加固法 soil cement processing, soil-ce-

ment processing
水泥土浆　cement-treated-soil grout
水平冻胀力　horizontal heave force
水平固结系数　coefficient of horizontal consolidation
水平滤井　horizontal filter well
水平位移计　horizontal movement gauge
水平钻孔排水　horizontal bore hole drain
水沙关系　relationship between water and sediment
水上沉桩法　pile sinking method on the water
水上打桩机　floating pile driver, pontoon pile driving plant
水头　hydrostatic head
水头损失　head loss, loss of head
水土保持　soil and water conservation, soil conservation
水土流失　loss of soil
水位　elevation of water
水位计　water level gauge
水位下降　drawdown
水位下降比　drawdown ratio
水位下降曲线　drawdown curve
水位骤降　rapid drawdown
水文地质勘察　hydrogeological investigation
水文地质学　hydrogeology
水文地质钻探　hydrogeological drilling
水文过程线形式　shape of hydrograph
水隙比　water void ratio
水下爆破　underwater blasting
水下打桩　underwater pile driving
水下工程　underwater work
水下基础　foundation under water
水下浇筑　placing under water

水下振砂器	terra-probe
水压测头	piezometer tip
水样	water sample
水云母	hydromica
水质分析	water quality analysis
水中称重法	weight in water method
水准测量	leveling, water level survey
水准点	bench mark
顺坝	longitudinal dike
顺层滑坡	consequent landslide
瞬时沉降	distortion settlement, initial settlement, undrained settlement
瞬时弹性应变	instantaneous elastic strain
瞬时荷载	instantaneous load, transient load
瞬时孔隙压力	instantaneous pore pressure
瞬时破坏	instantaneous rupture
瞬时压缩	immediate compression
斯宾赛法	Spencer method
斯肯普顿极限承载力公式	Skempton's ultimate bearing capacity formula
四分法	quartering
四片层矿物	four-sheet mineral
四通	double lee
松弛变量	slack variables
松弛函数	relaxation function
松弛模量	relaxation modulus
松弛时间	relaxation time
松弛试验	relaxation test
松弛效应	relaxation effect
松弛岩体	relaxed rock
松弛因数	relaxation factor
松动爆破	loose blasting
松动压力	loosening pressure

松铺厚度 lay-down thickness
松软土 mellow soil, mollisol
松软岩体 soft rock mass
松散冲积层 loose alluvium
松散地基 loose foundation
松散堆积物 rickle
松散土 loose soil, friable soil
松散岩石 friable rock
松散岩体 loose rock mass
松散岩土 non-cohesive soil
松砂 loose sand, open sand
松土机 pneumatic pick
松土搅拌机 pulvi-mixer
松土器 ripper equipment
送桩 follower, long dolly, pile follower
送桩锤 beetle head
送桩器 chaser
素混凝土 plain concrete
素填土 plain fill
素土垫层 plain soil cushion
素土夯实 rammed-earth
速凝剂 rapid setting admixture
塑料板排水 plastic drain
塑料板排水法 plastic board drain method
塑料带排水 plastic band-shaped drain
塑料滤布 plastic filter cloth
塑料膜 plastic membrane
塑料排水（带法） prefabricated strip drain, geodrain
塑料排水管 plastic drainage pipe
塑流 plastic flow
塑流区 zone of plastic flow
塑流现象 flow phenomenon
塑限 limit of plasticity, plastic limit

塑限试验	plastic limit test
塑限下限	lower plastic limit
塑性	plasticity
塑性本构关系	plastic constitutive relations
塑性变形	plastic deformation
塑性范围	plastic range
塑性分析法	plastic analysis method
塑性极限分析	plastic limit analysis
塑性极限分析定理	plastic collapse-basic theorem
塑性极限荷载	plastic limit load
塑性极限弯矩	plastic limit bending moment
塑性极限状态	plastic limit state
塑性计	plastometer
塑性铰	plastic hinge
塑性截面矩	section plastic modulus
塑性流动法则	plastic flow law
塑性流动曲线	plastic flow curve
塑性模量	modulus of plasticity, plastic modulus
塑性黏土	plastic clay
塑性-黏滞流动	plastic-viscous flow
塑性平衡	plastic equilibrium
塑性平衡状态	state of plastic equilibrium
塑性破坏	plastic failure
塑性破坏理论	plastic theory of failure
塑性强度	plastic strength
塑性区	plastic zone
塑性区半径	radius of plastic zone
塑性区最大深度	maximum depth of plastic zone
塑性失稳	plastic instability
塑性势	plastic potential
塑性试针	plasticity needle

塑性体积应变　plastic volumetric strain
塑性图　plasticity chart
塑性土　plastic soil
塑性稳定性　plasticity stability
塑性应变　plastic strain
塑性应变增量　increment of plastic strain
塑性指数　plastic index of clay, plasticity index
塑性状态　plastic state
酸度试验　acidity test
酸化　acidizing
酸碱度试验　acidity and alkalinity test
酸性土　acid soil
酸性岩　acidite
随机荷载　random load
随机排列　random arrangement
随机样品　chance sample
随挖随填　cut-and-cover
遂道开挖　tunnel excavation
遂洞导流　tunnel diversion
碎砾石　broken gravel, crushed gravel
碎砾岩　psephite
碎料　crushed aggregate
碎裂带　shattered zone
碎漂石　crushed boulder
碎石　break stone, broken stone, channery, crushed rock, crushed stone, detritus
碎石垫层　broken stone base course
碎石基础　macadam foundation
碎石路　macadam
碎石路面　macadam pavement
碎石土　stone
碎石土地基　crushed stone soil foundation, gravelly soil foundation

碎石桩　gravel pile, stone column
碎屑　fragment
碎屑沉积　clastic deposit, detrital sediment
碎屑岩　clasolite
隧道　tunnel
隧道测量　tunnel survey
隧道衬砌　tunnel lining
隧道导坑　tunnel heading
隧道底拱　tunnel invert
隧道顶板　tunnel roof
隧道洞蚀　tunnel erosion
隧道盾构　tunnel shield
隧道盾构施工法　shield tunnelling method
隧道防水　waterproofing of tunnel
隧道工程　tunneling
隧道截面　tunnel profile
隧道掘进机　tunnel borer, tunnel boring machine
隧道掘进机法　tunnel boring machine method
隧道排水　tunnel drainage
隧道弃方　tunnel spoil
隧道圈　tunnel ring
隧道施工　tunnel construction
隧道施工排水　tunnel drainage during construction
隧道竖井　tunnel shaft
隧道效应　tunnel effect
隧道凿岩机　tunnel drill
隧道支撑　tunnel support
隧洞入口　portal of tunnel
燧石　flint
损坏指数　damage index
缩比例　scaling factor
缩沉　shrink mark

缩颈 waist, waisting, gapping
缩限 shrinkage limit
缩限试验 shrinkage limit test
缩性指数 shrinkage index
锁紧力 locking force

T

塌落拱 ground arch
塌陷地震 collapse earthquake
塌陷坑 subsidence trough
踏勘阶段 reconnaissance phase
台阶式基础 benched foundation
台阶式开采 bench stoping
太古代 archaeozoic era
太沙基承载力理论 Terzaghi bearing capacity theory
太沙基地基极限承载力公式 Terzaghi's ultimate bearing capacity equation
太沙基固结理论 Terzaghi's consolidation theory
太沙基理论 Terzaghi theory
太沙基-伦杜列克扩散方程 Terzaghi-Rendulic diffusion equation
泰勒法 Taylor method
坍岸 bank caving
坍落 caving
坍落稠度试验 slump consistency test
坍落度 slump, slump constant
坍落度法 slump test
坍落度损失 slump loss
坍落度筒 slump cone

坍塌　slough
坍塌块体　slump block
坍塌页岩　sloughing shale
坍陷　collapse, collapse settlement
滩沉积　beach deposit
探槽　exploration trench, test trench
探槽取样　pit sampling
探测　probing
探洞　trial heading
探井　exploratory shaft, sounding well, test pit
探坑　exploratory pit, inspection pit, prospect pit, trial hole
探索性研究　exploration investigation
探头　probe
探针检验　feeler inspection
碳酸岩　carbonatite
碳质沉积　carbonaceous sediment
掏槽纠偏　tilting correction by earth undercutting
陶瓷滤头　ceramic filter
陶瓷泥浆　slurry of ceramic
陶管　ceramics
淘洗试验　elutriation test
套管　bore casing, casing, casing tubes
套管传动联接器　casing drive adapter
套管法灌浆　sleeve pipe grouting
套管混凝土桩　cased concrete pile
套管接触　casing collar
套管式灌浆法　telescope grouting method
套管头气体　casing head gas
套管靴　casing shoe
套管桩　cased pile
套管钻孔　cased borehole
套管钻孔桩　cased bore pile

套索桩　belaying pin
套桩　lag pile
套钻　overcoring
特大洪水　eventual flood
特雷斯卡屈服条件　Tresca yield condition
特殊基础　special foundation
特殊土　special soil
特征紊流　signature turbulence
梯度　gradient
梯度比试验　gradient ratio test
梯式加筋锚定墙　ladder wall, multi-anchored wall
梯状挡土墙　terrace wall
体（积）膨胀　cubic dilatation, cubical expansion
体比重　bulk specific gravity
体变潜量　potential volume change (PVC)
体波　body wave
体积百分比　percentage by volume
体积比热　volumetric specific heat
体积变形模量　volumetric deformation modulus
体积单位重量　bulk unit weight
体积力　mass force
体积模量（M_b）　bulk modulus
体积配筋率　reinforcement ratio per unit volume
体积热容量　volumetric heat capacity
体积系数　volumetric coefficient
体积压缩系数　coefficient of volume compressibility
体积压缩性　bulk compressibility
体密度　bulk density
体缩率　volume shrinkage ratio, volumetric

shrinkage
体应变　volumetric strain
体应力　body force
体胀　bulking
体胀率　volume change, volume expansion
替代方案　alternative scheme
天然稠度试验　natural consistency test
天然地层　natural strata
天然地基　subsoil, natural base, natural foundation, natural ground
天然地基允许荷载　subsoil permissible load
天然冻结法　natural freezing method, natural frozen method
天然洞穴　cavity
天然拱　natural arch
天然含水量　moisture content of natural, natural moisture content, natural water content
天然级配砾石　pit-run gravel
天然建筑材料　natural building materials
天然孔隙比　natural void ratio
天然排水　natural drainage
天然坡　natural slope
天然软黏土　natural soft clay
天然砂　natural sand
天然湿度　field moisture
天然土　native soil, natural soil
天然卸载拱　natural load-transmitting arch
天然休止角　natural angle of repose
填方　earth fill, fill, made ground
填方量　bank meassure
填海工程　reclamation works
填海土地　reclaimed land
填料　packing, padding

填塞 tamp
填塞材料 blinding material
填石盲沟 rock drain
填土 filled soil, landfill
填土沉陷 settlement of fill, sinking of filling
填土地基 fill(ed) ground
填筑坝 embankment dam
填筑高程 reclamation level
填筑含水量 placement water content
填筑湿度 placement moisture
填筑顺序 sequence of filling
填筑质量控制 quality control of earth-rock fill
条分法 finite slice method, method of slice, slice method
条形荷载 strip load
条形基础 strap footing, strip footing, strip foundation
铁路 rail facility
铁路道口 railway crossing
铁路涵洞 railway culvert
铁路路基 railroad bed
铁路桥 railway bridge
铁路隧道 railroad tunnel
铁遂石 taconite
铁质胶结物 ferruginous cement
停泊处 berth
停打阻力 home
通风平巷 ventilation lateral
通气带 aeration zone
同晶置换 isomorphous substitution
同位素 isotope
同位素反射密度仪 back scatter densimeter
同位素含水量探测仪 moisture probe
同质同晶 allomorph

同轴喷嘴　coaxial nozzles
统计水文学　statistical hydrology
统计土力学　statistic soil mechanics
统计损伤理论　statistical damage theory
统一土壤分类法　unified soil classification system
筒形结构　tubular construction
透镜体　lens
透气性　air permeability
透气性测定仪　permeability meter
透气性试验　air permeability test
透水材料　pervious materials, seepy material
透水层　permeable layer, pervious bed
透水地基　pervious foundation
透水垫层　pervious blanket
透水缝　permeable joint
透水铺盖　permeable blanket
透水铺盖层　blanket course
透水石　perforated stone, porous stone
透水性　permeability to water, hydraulic conductivity
透水性试验　water permeability test
透水岩层　permeable rock
透水岩石　porous rock
凸度　convexity
图段　segment
图解分析　graphical analysis
图示法　graphical representation
涂面桩　coated pile
土　earth, soil
土坝　earth dam
土坝护面　earth dam paving
土坝老化　earth dam ageing
土坝压实　earth dam compaction

土崩　earth fall, earth slide
土层　soil strata
土层反应计算　soil layer response calculation
土层厚度　the thickness of the layer
土层滑动　earth slip
土层锚杆　soil anchor
土层剖面　soil section
土承载力值　soil bearing value
土触探装置　soil sounding device
土袋埝坝　bag dam
土袋围堰　soil sack cofferdam
土的饱和度　saturation of soil
土的饱和密度　saturation density of soil
土的饱和容重　saturated unit weight of soil, saturation unit weight of soil
土的饱和重度　saturation weight density of soil
土的本构定律　constitutive law of soil
土的本构模型　constitutive model of soil
土的崩解性　slaking of soil
土的表层　mantle of soil
土的泊松比　Poisson's ratio of soil
土的不规则结构　erratic soil structure
土的层理　soil stratification
土的承压力　bearing force of soie
土的承载能力　soil bearing capacity
土的稠度　consistency of soil
土的次弹性模型　hypoelastic model of soil
土的弹塑性　elastic-plasticity of soil
土的弹塑性模型　theory of elastoplastic model of soil
土的弹性模量　elastic modulus of soil
土的弹性模型　elastic model of soil
土的动剪变模量　dynamic shear modulus of soil

土的动力性质　dynamic properties of soils
土的动力性质参数　dynamic property parameter of soil
土的动强度　dynamic strength of soils
土的分类　classification of soil
土的浮表观密度(浮容重)　submerged unit weight of soil
土的浮密度　submerged density of soil
土的浮重度　submerged weight density of soil
土的构造　soil texture
土的固结　consolidation of soil
土的固结理论　consolidation theory of soil
土的管涌　soil piping
土的加工硬化理论　theory of strain gardening law of soil
土的加固　soil strengthening
土的加筋法　soil reinforcement
土的剪切波速　shear wave velocity of soil
土的简易分类法　quick soil classification, rapid soil classification
土的胶结物　soil binder
土的结构　soil fabric, soil structure
土的结构强度　structural strength of soil
土的抗剪强度　shear strength of soil
土的抗剪强度参数　shear strength parameter of soil
土的颗粒级配　particle-size-distribution of soil
土的孔隙比　pore ratio of soil
土的灵敏度　sensitivity of soil
土的流动规则理论　theory of flow rule of soil
土的密度　density of soil
土的密实度　compactness of soil
土的膨胀变形量　deformation of expansion of soil

土的膨胀率 rate of swelling of soil
土的切线模量 tangent modulus of soil
土的屈服面理论 theory of yield surface of soil
土的肉眼分类 visual soil classification
土的渗透性 permeability of soil, soil permeability
土的识别 identification of soils
土的收缩变形量 deformation due to shrinkage of soil
土的收缩系数 coefficient of shrinkage of soil
土的水压力 soil water tension
土的塑性 plasticity of soil
土的体积压缩系数 coefficient of volume compressibility of soil
土的图例 soil legend
土的吸力 soil suction
土的细粒部分 soil fines
土的现场鉴别 field identification of soil
土的线弹性模型 linear elastic model of soil
土的相对密度 relative density of soil
土的性质 soil properties
土的压实 compaction of soil, soil compaction
土的压缩曲线 compression curve of soil
土的压缩系数 coefficient of compressibility of soil
土的压缩性 compressibility of soil, compression of soil
土的液化 soil liquefaction
土的有效密度 effective density of soil
土的原位试验 in-situ soil testing
土的振动压密 dynamic densification of soils
土的滞后弹性模型 hysteretic elastic model of soil
土的重度 weight density of soil

土的阻尼　soil damping
土的阻尼比　damping ratio of soil
土的组成　soil composition
土堤　earth embankment, soil bank
土堤岸护墙　chemise
土地平整　land-leveling
土钉　soil nailing
土钉墙　soil nailing wall
土动力学　soil dynamics
土方工程　earth work
土方工程量　volume of earthwork
土分布图　engineering soil map, soil map
土工布，土工织物　geotechnical fabrics (geofabric)
土工布挡土墙　fabric retaining wall
土工布反滤层　filter fabric mat
土工布反滤层试验　filter fabric soil retention test
土工参数　soil parameter
土工垫　geomat
土工复合材料　geocomposite
土工格栅　geogrid
土工工程　earthwork engineering
土工合成材料　geosynthetics
土工结构物　earth structure
土工聚合物　geopolymer
土工离心模型试验　geotechnical centrifugal model test
土工模袋　geofabriform
土工模型试验　geotechnical model test
土工膜　geomembrane
土工试验　soil test
土工试验室　soil lab
土工网　geonets
土工织物　geotextiles

土拱效应　soil arch effect
土骨架　soil skeleton
土固结　soil consolidation
土滑动面　rupture plane of slope
土及岩石试验　soil and rock testing
土建筑　rammed earth building
土胶体　soil colloid
土-结构相互作用　soil-structure interaction
土孔隙率　soil porosity
土块实验　clod test
土矿物学　soil mineralogy
土类　soil group
土力学　soil mechanics
土粒　soil grain, soil particle
土粒密度　density of solid particles
土粒相对密度（比重）　specific density of solid particles, specific gravity of soil particle
土梁试验　earth beam test
土料　earth material
土锚杆　earth anchor
土名　soil name
土内等孔隙水压线　soil lines of equal pore pressure
土坯　adobe
土坡　slope, soil slope
土坡滑塌　failure of earth slope
土坡基底破坏　base failure of slope
土坡临界高度　critical height of slope
土坡稳定分析　stability analysis of slope
土坡稳定分析的总应力法　total stress method of stability analysis
土壤持水曲线　soil water retention curve
土壤防冻　soil freezing prevention

土壤分类　soil classification
土壤抗陷系数　depressive coefficient of soil
土壤空气　soil air
土壤空隙　interstices of soil
土壤粒径分布试验　soil grain distribution test
土壤密实度　soil compactibility
土壤水　holard
土壤水分特征　soil water retension characteritics
土壤松散度　looseness of soil
土壤松散系数　loose coefficient of soil
土壤学　agrology, edaphology, pedology
土壤蒸发　soil evaporation
土壤自然排水　natural soil drainage
土壤自然坡度角　natural slope angle of soil
土塞作用　plug effect
土勺灌入试验　spoon penetration test
土石坝　earth-rock dam
土石场现场鉴别法　on-site identification of soil and rock
土石方工程　earthwork
土石方开挖　earth-rock excavation
土体　soil mass
土体变形　deformation of soil mass
土体的残余变形　residual deformation of soil mass
土体的触变　thixotropy of soil mass
土体的剪胀　dilatation of soil mass
土体固结　consolidation of soil mass
土体加固法　soil stabilization
土体剖面　soil profile
土体压后沉降量　settlement of compacted soil
土围堰　earth cofferdam, soil cofferdam
土吸力势　soil suction potential

土系　soil series
土楔　soil wedge
土选分机　soil separator
土压力　earth pressure, soil pressure
土压力分布　earth pressure distribution
土压力盒　earth pressure cell
土压力系数　earth pressure coefficient
土压平衡式盾构　earth pressure balanced shield, shield with balanced earth pressure
土样　soil pattern, soil sample
土样采取率　percent of sample recovery
土样分析筛　soil sample analysis sieve
土样钻取器　soil auger
土液化机理　mechanism for soil liquefaction
土质地基　earth foundation
土质调查　soil survey
土质改良　soil improvement
土质勘探　soil exploration
土中气　air in soil
土中水　soil moisture, soil water, water in soil
土中水分应力　soil moisture stress
土中水分运动　soil water movement
土中应力波　stress wave in soils
土桩　earth pile, soil column, soil compaction pile, soil pile
土桩加固法　soil column stabilization method
土钻　earth borer, earth drill
湍流　turbulence
团粒　cumularspharolith
团粒结构　cluster structure, crumb-structure, aggregate structure
推力　thrust, thrust force

推力曲线　thrust curve
推土机　bulldozer, dozer, soil shifter
推样土器　sample extrude
托换技术　underpinning technique
托换桩　underpinned pile
脱水试验　dehydration test
拓宽　frontiers

W

挖出物　dredged spoils
挖方　excavated volume, excavation, cut
挖方边坡　site slope of excavation
挖方坡度　excavation slope
挖方填方　cut-and-fill
挖沟法　trench cut method
挖掘机　navvy
挖掘机械　excavating machinery
挖坑沉基　foundation by pit sinking
挖孔灌注桩　dug cast-in-place pile, manually excavated cast-in-place pile
挖孔桩　cast-in-situ pile by excavation, excavated pile
挖泥船　drag boat, dredger
挖土机　excavator
蛙式打夯机　frog-rammer
外附力　stick force
外加力　applied fore
外间隙比　outside clearance ratio
外力　external force
外摩擦角　angle of external friction
弯剪裂缝　shear-bending crack
弯矩　bending moment

弯矩应力函数　stress function of bending
弯矩再分配　redistribution of moment
弯曲应力　bending stress
弯液面　meniscus
弯状取芯钻　calyx drill
弯状钻孔　calyx drill holes
完全固结　thorough consolidation
完整井　completely penetrated well
完整性检测　integrity testing
完整岩石　intact rock
晚第三纪　Neogene period
网格式盾构　shield faced with grid
网格组织　cellular texture
网络树根桩　reticulated root piles
网式砌墙　reticulated masonry
网状结构　reticular structure
往返活动性　reciprocating activity
往复直剪试验　reversing shear box test
微波含水量仪　microwave moisture apparatus
微观结构　microstructure
微观组织　micro texture, micro-texture
微化石　microfossil
微粒灌浆　particulate grouting
微裂缝　tiny crack
微裂纹　micro-crack
微裂隙　microfissure
微生物处理　microbiological treatment
微团粒　micro-aggregate
微型桩　mini pile
微震　microseism
微组构　microfabric
围垦工程　reclamation
围压　confining pressure
围岩　adjacent rock, adjoining rock, surround-

ing rock, wall rock
围岩变形观测 observation of surrounding rock deformation
围岩变形压力 deformational pressure of surrounding rock
围岩弹性变形压力 elastic deformational pressure of surrounding rock
围岩流变压 rheological pressure of surrounding rock
围岩松动区 loosening zone of surrounding rock
围岩松动压力 loose pressure of surrounding rock
围岩位移 displacement of surrounding rock
围岩稳定性 stability of surrounding rock
围岩压力 pressure of surrounding rock, surrounding rock pressure
围岩应力 stress in the surrounding rock, surrounding rock stress
围堰 cofferdam
围堰支撑 cofferdam-bracing
帷幕灌浆 curtain grouting
维持荷载法 maintained load test
尾矿坝 mine tailings dam, tailings dam
尾矿砂 tailings
未变质岩石 unaltered rock
未冻水量 unfrozen water content
未结合水 uncombined water
未扰动地基 intact ground
未扰动土地 undisturbed ground
位密度 bit density
位移 displacement
魏汝龙-Khosla-Wu 模型 Wei Rulong-Khosla-Wu model

魏锡克极限承载力公式	Vesic's ultimate bearing capacity formula
温差应力	stress due to temperature difference
温度裂缝	thermal cracking
温度作用	temperature action
文克尔地基模型	Winkler foundation model
文克勒地基	Winkler foundation
文克勒假定	Winkler's assumption
纹理	texture
紊流	turbulent flow
稳定(性)系数	stability factor
稳定安全系数	safety factor of stability
稳定方程	stability equation
稳定分析	stability analysis
稳定河床	permanent bed
稳定极限	limit of stability
稳定浆液	stabilizing grout
稳定理论	theory of stability
稳定力矩	stabilizing moment
稳定裂纹扩展	stable crack growth
稳定蠕变	steady creep
稳定渗流	steady seepage(flow)
稳定数	stability number
稳定水位	steady water table
稳定性	stability
稳定性加强系统	stability augmentation system
稳定性勘察	stability investigation
稳定性评估	stability assessment
稳定性曲线	stability curve
稳定岩基	stable rock bed
稳态流	stationary flow, stationary stream
稳态流场	stationary flow field
稳压器	manostat

沃斯列夫参数　Hvorslev parameter
沃斯列夫面　Hvorslev surface
沃斯列夫模型　Hvorslev soil model
渥太华砂　Ottawa sand
圬土基础　foundation of masonry
无侧限抗压强度　unconfined compression strength
无侧限抗压强度试验　unconfined compressive strength test
无侧限压缩试验　non-confined compression test, unconfined compression test
无侧向应变试验　zero lateral strain test
无掺料水泥　simple cement
无定形物质　amorphous substance
无纺土工织物　nonwoven geotextile
无机土　inorganic soil
无黏性土　cohesionless soil, frictional soil, non cohesive soil
无气水　air free water
无声打桩机　silent pile driver
无塑性土　non plastic soil
无损探测技术　non-destructive techniques
无损探伤　non-destructive detection
无套管桩　uncased pile
无套管钻探　open-hole drilling
无限边坡　infinite slope
无限元法　infinite element method
无效爆破　spent shot
无压流　non-pressure flow
无岩芯钻进　noncore drilling
无载饱和曲线　no-load saturation curve
无震区　aseismic region
无支撑挖掘　unbraced excavation

无桩靴夯扩灌注桩　rammed bulb pile
物理风化　physical weathering
物理模型　physical model
物理相互作用　physical interaction
物理相似　physical similitude
物探　geophysical prospecting
误差传播　propagation of error
误差限度　limit of error

X

吸附层　absorbed layer, adsorbed layer, adsorption layer
吸附界限　absorption limit
吸附能力　absorptive capacity
吸附水　adsorbed water, hydration water, hydroscopic water, planar water, retained water
吸附水膜　capillary film
吸附损失　absorption loss
吸附压力　adsorbent pressure
吸混作用　persorption
吸浆量　acceptance of grout, grout absorption, grout acceptance, rate of grout acceptance
吸浆率　acceptance rate
吸泥下沉沉井　open caisson sinking by suction dredge
吸热反应　endothermic reaction
吸湿含水量　hygroscopic water content
吸湿容量　hygroscopic capacity
吸湿系数　hygroscopic coefficient
吸湿性　hydroscopic moisture

吸收率　absorptance
吸水井　section well, absorbing well
吸水试验　bleed test
吸着水　absorbed water
希利公式　Hiley formula
矽质黏土岩　argillite
系缆柱　dolphin
细长柱　slender column
细度　degree of fineness
细度模数　fineness modulus
细粉粒粘结　fine silt bond
细粉土　fine silt
细级配的　fine-graded
细颗粒分析　fine analysis
细粒含量百分率　percent fines
细粒料　fine aggregate
细粒料填缝　seal with fines
细粒土　fine grained soil, fines, fine-grained soil
细砂　fine sand, graining sand
峡谷　gulch
下沉　subsidence
下降漏斗　cone of depression
下孔法　down-hole method, down-hole shooting
下拉荷载　downdrag
下卧层　substratum, underlying stratum
下卧土　underlying soil
下限定理　lower bound theorem
下游法　downstream method
先期固结　preconsolidation
先期固结土　preconsolidated soil
先期固结压力　preconsolidation pressure
先张拉法　stressing method

纤维加强混凝土　fiber-reinforced concrete
纤维素纤维　cellulosic fibers
纤维填充物　fiberfill
纤维土　texsol
纤维性泥炭　fibrous peat
纤维性土　fibrous soil
现场拌合　Mixed-in-place
现场爆炸试验　field blasting test
现场测试　field measurements
现场抽水试验　field pumping test
现场触探试验　field sounding test
现场调查　on-site investigation
现场工作　on-site work
现场观测　field observation
现场含水当量　field moisture equivalent
现场监测　in-situ monitoring
现场剪切触探仪　iskymeter
现场鉴定　field identification
现场浇筑混凝土　poured-in-place concrete
现场勘察　field investigation
现场控制　field control
现场密度试验　field density test
现场碾压试验　field compaction test
现场强度　in-situ strength
现场渗透试验　field permeability test
现场十字板试验　field vane test
现场试验　test on site
现场试验室　field laboratory, field test
现场数据　field data
现场踏勘　site survey
现场土　soil in place, soil in situ
现场压实曲线　field compaction curve
现场载荷试验　field bearing test, field loading test

现场再压缩曲线　field recompression curve
现场作业　field operation
现代冰川作用　contemporary glaciation
现浇薄膜防渗墙　thin cast-in-situ diaphragm wall
现浇混凝土桩　in-situ concrete pile
现行标准　actual standard
线弹性分析法　linear elastic analysis method
线荷载　line load, linear load
线缩率　linear shrinkage ratio
线位移　linear displacement
线性　linearity
线性变形模量　modulus of linear deformation
线性可变差动变压器　linear variable differential transformer
线性黏弹性模型　linear viscoelastic model
线性膨胀系数　coefficient of linear expansion
线应变　linear strain
线应变率　linear strain ratio
线胀率　linear expansion
限差　tolerance
限制粒径　constrained diameter(grain size)
相对沉降量　relative settlement
相对高程　relative elevation
相对含冰量　relative ice content
相对埋深　relative embedment
相对密度（比重）　relative density, specific density, specific gravity
相对密度试验　relative density test
相对密实度　relative compaction
相对膨胀　relative expansion
相对湿度　relative humidity
相对受压区高度　relative depth of compressive area

相对位移 relative displacement
相对误差 relative error
相对硬度 relative hardness
相关系数 correlation coefficient
相互影响系数 interactive coefficient
相互作用 interaction effect
相角 phase angle
相邻槽段 adjacent panels
相平面 phase-plane
相容方程 compatibility equation
相适应流动法则 associative flow rule
相似定律 law of similarity
相似理论 principal of similitude
相似系数 scale factor
相位差 phase difference
箱格形心墙 cellular core wall
箱形钢桩 box pile
箱形基础 box foundation
镶边石 border stone
详细勘探 detailed exploration
向斜 syncline
向源浸蚀 backward erosion
橡皮膜 membrane
橡皮膜嵌入效应 membrane penetration effect
橡皮膜顺变性 membrane compliance
橡皮膜校正 membrane correction
削石坡 rock cut slope
消除应力 release of stress
消极隔振 passive isolation
消融 ablation
消融碛 ablation moraine
消散 dissipation
消散试验 dissipation test
消声打桩法 muffler piling

消声罩	acoustic box
消声作用	acoustic blanking
小砾石	pea gravel
小山坡	hillside
小石块	finger stone
小型夯土机	small-size mechanical rammer
小型十字板仪	pocket shear meter
小型桩	micro-pile
小应变	small strain
小主平面	minor principal plane
小主应变	minor principal strain
小主应力	minor principal stress
效率公式	efficiency formula
效率系数	efficiency factor
楔入效应	wedging effect
楔体理论	wedge theory
楔形掏槽	V-shape cut
楔形掏槽	wedge cut
歇后增长	freeze
斜层理	diagonal bedding
斜长石	anorthose
斜撑	batter brace, knee brace
斜撑加强支架	reinforced stulls
斜导架式打桩机	batter leader pile driver
斜断层	diagonal fault
斜角板桩	bevel sheet pile
斜节理	diagonal joint, oblique joint
斜截面承载力	shear capacity of inclined section
斜截面抗剪强度	shear strength of inclined section
斜井	inclined shaft
斜井地压	rock pressure in inclined shaft
斜坡崩塌	slope failure

斜坡排水　batter drainage
斜坡坡度　ramp slope
斜坡稳定性　slope stability
斜坡稳定性分析　slope stability analysis
斜坡整治工程　slope treatment works
斜墙　sloping core
斜纹受剪　shear slant to grain
斜桩　angle pile, batter pile, inclined pile, racking pile
携带式地下水位测定仪　portable dipmeter
泄洪隧洞　spillway tunnel
泄水孔　weep hole
泻湖沉积　lagoonal deposit
卸荷　decompression
卸荷模量　decompression modulus
卸荷曲线　decompression curve
卸荷试验　unloading test
卸荷台　relief platform
卸压再压环　decompression and recompression loop
卸载　unloading
卸载模量　unloading modulus
卸载曲线　unloading curve
谢尔贝薄壁取样器　Shelby tube sampler
心墙　core wall
心墙堆石坝　core wall type rockfill dam
心墙截水槽　core trench
心墙土坝　core earth dam
芯棒　mandrel
辛普公式　Simpson formula
新奥法　New Austrian Tunnelling Method
新采砂　green sand
新残积土　immature residual soil
新成土　entisol

新黄土　neo-loess, young loess
新近沉积黏土　young clay
新近堆积土　recently deposit soil
新生代　Cainozoic era, Cenozoic era, Neozoic era
新石器时代　new stone age
新鲜岩石　fresh rock, unweathered rock
行业规范　codes of practice
形函数　shape function
性状　behaviour
胸墙　breast wall
休止角　angle of repose
修正格里菲斯准则　modified Griffith's criterion
修正麦卡利地震烈度表　modified Mercalli scale
袖珍贯入仪　pocket penetrometer
虚功原理　principal of virtual work
虚应力　virtual stress
徐变损失　loss of creeping
絮凝结构　flocculent structure
絮凝土　flocculent soil
絮凝作用　flocculation
絮状物　floc
絮状支托粘结　flocculated clay buttress bond
玄武岩　basalt
悬臂式板桩墙　cantilever sheet pile wall, cantilever sheet piling
悬臂式挡土墙　cantilever retaining wall
悬臂式基础　cantilever footing
悬挂式帷幕　hanging curtain
旋冲钻　churn drill
旋回沉积作用　cyclotheric sedimentation
旋流泵　cyclone pump
旋喷管　jet grouting battery
旋喷柱体　jet column

旋喷桩　chemical churning pile
旋喷桩柱　jet grouting column
雪崩　avalanche
循环风　recirculation of air
循环荷载下土的应力-应变关系
　　stress-strain relationship of soil under cyclic loading

Y

压变控制试验　controlled-strain test
压浆泵　mud jack
压浆法　mortar grouting method, mud jack method
压力　compression
压力泵　forcing pump
压力波　pressure wave
压力传感器　pressure sensor, pressure transducer
压力分布　pressure distribution
压力拱　pressure arch
压力灌浆　injection grout
压力灌浆法　pressure grouting
压力灌浆管　pressure grout pipe
压力灌注　pressure injection
压力灌注桩　pressure pile
压力盒　load cell, pressure cell
压力环　pressure ring
压力计　manometer, piezometer, pressure gauge
压力泡　pressure bulb
压力喷浆　pneumatic mortar
压力试验　pressure test
压力损失　loss of pressure

压力系数	pressure coefficient
压力下降	pressure drop
压力仪	pressiometer
压力枕	pressure cushion
压泥浆	mud jacking
压平	ironing
压坡	counterpoising
压强水头	pressure head
压曲	buckling
压曲荷载	buckling load
压入式取样器	pressed sampler
压入桩	jacked pile, pressed pile
压砂法	sand flow method
压实	compaction, compacting
压实遍数	compactor pass
压实参数	compaction parameter
压实度	degree of compaction
压实分层厚度	compacted lift
压实机	compactor
压实力	compactive effort
压实深度	compacted depth
压实填土	compacted fill, compacted soil
压实系数	coefficient of compaction, compaction factor, percent compaction, compacting factor
压实性	compactibility
压水试验	pump-in test, test by injecting water into the grout hole, water pressure test
压水试验记录	water test log
压水试验孔	water pressure test hole
压缩比	compressibility ratio, compression ration
压缩变形	compression deformation, compres-

sive deformation
压缩波 compression wave
压缩层厚度 thickness of compressed layer
压缩空气锤 compressed-air hammer
压缩空气夯 compressed-air tamper
压缩空气洗井钻进 air drilling
压缩量 amount of compression
压缩模量 constrained modulus, modulus of compressibility, modulus of compression
压缩破坏 fail in compression
压缩区 compression zone
压缩曲线 compression curve, pressure-void ratio curve
压缩试验 compression test
压缩系数 coefficient of compressibility, coefficient of compression
压缩性 compressibility
压缩性地基 compressibility foundation
压缩性土 compressible soil
压缩指数 coefficient index
压缩指数 compression index
压土器 soil packer
压应变 compressive strain, compression strain
压应力 compressive stress, compression stress
压载滤水体 loaded filter
压重 kentledge
压重填土 counterweight fill
压桩力测量装置 pile pressing force measuring device
压桩力传递系统 pile pressing force transmitting system
压桩速度 speed of pressuring pile
亚稳结构 metastable structure

亚稳土 metastable soils
延迟灌浆 deferred grouting
延伸 stretching
延时沉降 delayed settlement
延性系数 ductility factor
严重破坏 severe damage
岩坝 rock bar
岩爆 rock burst
岩崩 rock avalanche, rock slide, rock fall
岩层 rock bedding, rock formation, rock stratum, rock stratification
岩层面 formation level
岩层钻孔 rockhole
岩顶锚杆支撑 rock roof bolting
岩洞 rock house, abra
岩化作用 lithification
岩基灌浆 rock foundation grouting
岩基稳定性 stability of foundation rock
岩浆 magma
岩浆岩 magmatic rock
岩块式倾倒 rock block toppling
岩锚支护 rock-bolt supporting
岩面 rock surface
岩坡平面破坏分析 plane failure analysis of rock slope
岩坡稳定分析 stability analysis of rock slope
岩坡楔体破坏分析 wedge failure analysis of rock slope
岩溶地层 karstic formation
岩溶地形 karst topography
岩溶洞 karst cave
岩溶井 karst pit
岩溶漏斗 karst funnel
岩溶水 karst water

岩溶水文学　karstic hydrology
岩溶特征　karst feature
岩石　rock
岩石本构关系　constitutive relation of rock, constitutive relationship of rock mass
岩石变形模量　deformation modulus of rock
岩石标准试件　standard specimen for rocks
岩石残余应力　residual stress in rock
岩石插筋锚固　rock pin
岩石长期强度　long-time strength of rock
岩石常规三轴试验　conventional triaxial test for rocks
岩石常规三轴试验机岩芯钻探　conventional triaxial testing machine for rocks
岩石初始蠕变　primary creep of rock
岩石单轴压缩强度　uniaxial compression strength of rock
岩石单轴压缩应变　uniaxial compression strain of rock
岩石的崩解性　slaking characteristic of rock
岩石的变形能　deformational energy of rock
岩石的泊松比　Poisson's ratio of rock
岩石的弹性模量　elastic modulus of rock
岩石的弹性滞后　elastic hysteresis of rock
岩石的动泊松比　dynamic Poisson's ratio of rock
岩石的动力特性　dynamic properties of rock
岩石的剪切破坏　shear rupture of rock
岩石的耐崩解性指标　slake-durability index of rock
岩石的全孔隙度　total porosity of rock
岩石的时间效应　time-dependent effects of rock

岩石的水理性质　physical properties of rock under water
岩石的塑性　plasticity of rock
岩石的物理力学性质　physicomechnanical property of rock
岩石的物理性质　physical properties of rock
岩石的应变软化性　strain-softening behaviour of rock
岩石的应变硬化性　strain-hardening behaviour of rock
岩石地基　rock foundation
岩石动弹性模量　dynamic elastic modulus of rock
岩石动力学　rock mass dynamics
岩石断裂力学　rock fracture mechanics
岩石断裂韧度　fraction toughness of rock
岩石墩　rock mound
岩石分布图　solid map
岩石分级　rock rating
岩石分类　petrographic classification, rock classification
岩石分析　rock analysis
岩石风化　rock decay, rock weathering
岩石风化程度　degree of rock weathering, weathering degree of rock
岩石覆盖层　rock cover
岩石割线模量　secant modulus of rock
岩石工程　rock engineering
岩石构造　structure of rock, rock structure
岩石构造学　petrotectnonics
岩石固结　rock bonding
岩石灌浆　rock grouting
岩石基质压缩系数　rock matrix compressibility
岩石加速蠕变　accelerated creep of rock

岩石坚硬程度　hardness degree of rock
岩石间接拉伸试验　indirect tensile test for rocks
岩石剪切试验　shear test for rock
岩石节理　rock joint
岩石节理剪力试验　rock joint shear strength test
岩石结构　rock texture
岩石结构　texture of rock
岩石抗剪强度　shear strength of rock
岩石矿物　rock forming mineral
岩石扩容　dilatancy of rock
岩石类型　rock type
岩石(体)力学　rock mechanics
岩石立面　rock-face
岩石裂隙　rock fissure
岩石裂隙张开度　aperture of rock fissure
岩石露头　rock outcrop
岩石锚杆测力计　rock bolt dynamometer
岩石锚杆理论　rock bolting theory
岩石锚杆伸长仪　rock bolt extensometer
岩石锚杆(栓)　rock bolt, rock bolting
岩石扭转试验　torsion test for rock
岩石劈理　rock cleavage
岩石平巷　stone drift
岩石坡，石坡　rock slope
岩石坡稳定性　rock slope stability
岩石破裂后期模量　post-failure modulus of rock
岩石破碎强度　rock crushing strength
岩石强度　rock strength, strength of rock
岩石强度理论　strength theory of rock
岩石强度曲面　strength curve surface of rock
岩石强度损失率　strength loss ratio of rock

岩石切线模量　tangent modulus of rock
岩石屈服　rock yield
岩石圈　lithosphere
岩石蠕变　rock creep
岩石软化系数　softening coefficient of rock
岩石软化性　softening property of rock
岩石三轴抗压强度　triaxial compression strength of rock
岩石三轴压缩应变　triaxial compression strain of rock
岩石声发射　acoustic emission of rock
岩石试验　rock testing
岩石塑性破坏　plastic rupture of rock
岩石透水性　permeability of rock
岩石突出　rock bump
岩石稳定蠕变　steady-state creep of rock
岩石系数　rock mass factor
岩石细裂隙　fine rock fissure
岩石学　petrology, lithology
岩石压力迹象　rock pressure indications
岩石压力理论　rock pressure theory
岩石要素　rock element
岩石应力　rock stress
岩石应力松弛　relaxation of rock
岩石应力-应变后期曲线　post-failure stress-strain curve of rock
岩石预应力　rock prestressing
岩石预应力假说　rock prestressing theory
岩石真三轴试验　true triaxial test for rock
岩石真三轴试验机　true triaxial testing machine for rock
岩石质量系数　rock quality index
岩石质量指标　rock quality designation(RQD)
岩石组成　petrographic composition

岩石钻机　machine rock drill
岩体　rock mass
岩体变形机制　mechanism of rock mass deformation
岩体变形模量　deformation modulus of rock mass, deformation of rock mass
岩体不连续面　rock discontinuities
岩体的各向异性　anisotropy of rock mass
岩体工程质量指标　engineering quality index of rock mass
岩体基本质量　rock mass basic quality(BQ)
岩体结构类型　structural types of rock mass
岩体抗剪强度　shear strength of rock mass
岩体破坏　failure of rock mass
岩体强度　rock mass strength
岩体强度　strength of rock mass
岩体损伤　rock damage
岩体完整性指数　intactness index of rock mass
岩体稳定性　stability of rock mass
岩体应力测试　rock mass stress test
岩土　rock and soil
岩土工程(学)　geotechnical engineering
岩土工程技术　geotechnique
岩土工程测试　geotechnical engineering test
岩土工程方法　geotechnical processes
岩土工程分级　categorization of geotechnical projects
岩土工程勘察　geotechnical investigation
岩土工程勘探　geotechnical exploration
岩土工程师　geotechnician
岩土工程学　geotechnology
岩土灌浆　geotechnical grouting
岩土力学参数　rock and soil mechanical param-

eters
岩土室内试验 indoor test of geotechnique
岩土现场试验 in-situ test of geotechnique
岩土与地下工程 geotechnical engineering and underground engineering
岩土支撑面 soil/rock support
岩屑 debris
岩屑滑动 debris slide
岩芯回收率 core recovery
岩芯 bore core, core of rock, rock core, core
岩芯采取率 percentage of core recovery
岩芯管 barrel
岩芯切取筒 core cutter
岩芯取样器 core sampler
岩芯钻 core drill
岩芯钻机 core drill rig
岩芯钻孔 core hole
岩芯钻筒 core barrel
岩芯钻头 core bit, coring bit
岩穴 rock pocket
岩柱 rock pillar
岩组学 rock fabric
盐度 salinity
盐基交换 base exchange
盐基交换容量 base exchange capacity
盐碱土 halomorphic soil, saline-alkali soil
盐型絮凝 salt-type flocculation
盐渍土 saline soil, salty soil
衍射 diffraction
验收试验 acceptance test
验证性试验 proof rest
堰 weir
堰塞湖 barrier lake
阳离子 cation, kation

阳离子交换　cation exchange
阳离子交换活动性　cation exchange activity
阳离子交换量　cation exchange capacity [CEC]
洋流　ocean current
扬压力　uplift pressure
氧化　oxidation
氧化物　oxide
样本大小　sampling size
样本协方差　sample covariance
窑灰　kiln dust
遥感勘测　remote sensing prospecting
咬合作用　interlocking action
野外勘探　field exploration
野外作业　field work
叶理　foliation
曳阻力　drag force
页岩　shale
页岩陶粒　shale ceramicite
页岩质黏土　shaly clay
液化　fluidification, liquefaction
液化初步判别　preliminary discrimination of liquefaction
液化的再判别　secondary identification of liquefaction
液化地基加固　improvement of liquefiable soil
液化度　degree of liquefaction
液化破坏　liquefaction failure
液化区　liquefaction zone
液化砂土地基加固　consolidation of liquefiable sandy soil
液化势　liquefaction potential
液化应力比　stress ratio of liquefaction
液化指数　liquefaction index
液塑限联合测定仪　liquid-plastic limit com-

bined device
液态　liquid state
液限　limit of liquidity, liquid limit
液限试验　liquid limit test
液限仪　liquid limit apparatus, liquid limit device
液相　liquid phase
液性指数　liquidity index, relative water content, water-plasticity ratio
液压千斤顶　hydraulic jack
一般堆积土　ordinary deposited soil
一点应变状态　strain state at a point
一点应力状态　stress state at a point
一维固结　one dimensional consolidation
伊利石　illite
医疗匣　medical lock
仪器埋设　internal instrument installation
仪器误差　instrumental error
移液管法　pipette method
异常水位　abnormal water level
异形灌注桩　special-shaped cast-in-place pile
易冻土　frost-susceptible soil
易冻性　frost susceptibility
易裂性　fissibility
溢水式沉降计　hydraulic overflow settlement cell
翼墙　abutment wall
阴极保护　cathodic protection
阴离子　anion
阴离子交换　anion exchange
阴离子吸附　anion adsorption
引起应力的外力　stress-producing force
引水隧洞　water intake tunnel
引线　fuse
隐蔽工程，埋设仪器　embedded construction

隐晶结构(沉积岩)　aphanocrystalline texture
隐晶结构(火成岩)　aphanitic texture
应变杆　tell-tale
应变花式应变仪　strain rosette
应变计　strain cell
应变开裂　strain cracking
应变空间　strain space
应变控制　strain control
应变控制式三轴压缩仪　strain control triaxial compression apparatus
应变控制试验　strain controlled test
应变裂缝　strain crack
应变率　strain rate, strain ratio
应变能　strain energy
应变偏斜张量　strain deviator tensor
应变强化阶段　strain-hardening stage
应变球形张量　strain spherical tensor
应变曲面　strain surface
应变曲线　strain curve
应变软化　deformation softening, strain softening
应变时效　strain aging
应变式位移传感器　strainer-gage type displacement transducer
应变图　strain diagram
应变向量　strain vector
应变协调因子　strain coordination factor
应变仪　strain gauge, strain meter
应变硬化　strain harding
应变张量　strain tensor
应变滞后　strain lag of added concrete
应急措施　emergency measures
应急灌浆　emergency grouting
应急井　emergency shaft

应力　stress
应力比　specific value of stress, stress ratio
应力比法　stress-ratio method
应力变形特征　stress-deformation characteristic
应力场　stress field
应力超前　stress excess of tensile steel
应力调整　stress regulation
应力冻结效应　stress-freezing effect
应力反复　stress repetition
应力范围　stress range
应力分布　stress distribution
应力分布图　stress diagram
应力分析　stress analysis
应力腐蚀　stress corrosion
应力腐蚀开裂　stress corrosion cracking
应力函数　stress function
应力恢复法　stress recovery method
应力及应变反应　stress and strain response
应力集中　stress concentration
应力集中的扩散　stress concentration diffusion
应力集中系数　stress concentration factor
应力计　stress gauge
应力解除　stress relieving, stress release
应力解除槽　stress-release channel
应力解除法　stress relief method
应力解除区　destressed zone, distressed zone
应力解除钻孔　stress releasing borehole
应力解法　stress method
应力空间　stress space
应力空间转换　transformation of stress space
应力控制式三轴压缩仪　stress control triaxial compression apparatus
应力控制试验　controlled-stress test, stress controlled test

应力扩散　stress dispersion
应力历史　stress history
应力裂缝　stress crack
应力路径　stress path
应力路径法　stress path method
应力蒙皮作用　stressed skin action
应力偏斜张量　stress deviatoric tensor
应力平衡法　stress-equilibrium method
应力谱　stress spectrum
应力强度因子　stress strength factor, stress intensity factor
应力强度因子计算方法　stress intensity factor calculation method
应力球形张量　stress spherical tensor
应力曲面　stress surface
应力水平　stress level
应力松弛　relaxation of stress, stress relaxation
应力松弛仪　stress relax meter
应力梯度　stress gradient
应力向量　stress vector
应力引起的断裂　stress induced fracture
应力应变比　stress to strain ratio
应力-应变曲线　stress-strain curve
应力-应变图　stress-strain diagram
应力圆　circle of stress, stress circle
应力约束　stress constraint
应力张量　stress tensor
应力张量不变量　invariants of stress tensor
应力重分布　stress redistribution
应力状态　state of stress
影响半径　radius of influence
影响漏斗　cone of influence
影响深度　depth of influence

硬表层	crust
硬度	hardness
硬度比	solidity ratio
硬度分类	class of hardness
硬化	hardening
硬化参数	hardening parameter
硬化规则	hardening rule
硬化模量	harden modulus
硬壳	overlying crust
硬黏土	firm clay, hard clay, leck, stiff clay
硬盘(土)层	duripan
硬石膏	karstenite
硬水	hard water
硬土	pan soil
硬土层	hardpan, orterde, pan formation
硬质黏土	flint clay
永冻层	ever frozen layer
永冻层减薄	permafrost degradation
永冻土	ever frozen soil, permafrost
永冻土面	permafrost table
永冻土增厚	permafrost aggradation
永久变形	permanent deformation
永久冻土	perpetually frozen soil
永久荷载	permanent load
永久性加固	permanent strengthening
永久支护	permanent lining
涌入	blow-in
涌水量	water inflow
用盾构法掘进隧道	shield-driven tunneling
优化模型	optimization model
优化设计	optimization for design
油漆打底	paint base
油漆底层	paint filler
有害沉降	detrimental settlement

有机玻璃　perspex
有机沉淀物　organogenous sediments
有机化合物　organic compound
有机环境　organic environment
有机离子　organic ion
有机黏土　organic clay
有机土　Histosol
有机质　organic matter
有机质含量　organic content
有机质土　organic soil
有锚定墙支撑的板桩堤岸　sheet-pile bulkhead with anchor-wall support
有套管的混凝土挡土墙　shelled concrete pile
有限差分法　finite difference method
有限元法　finite element method [FEM]
有效沉降量　effective deformation
有效法向应力　effective normal stress
有效固结应力　effective consolidation stress
有效截面积　effective cross section
有效孔隙率　effective porosity
有效粒径　effective grain size, effective diameter, effective size
有效内摩擦角　effective angle of internal friction
有效筛孔　effective opening
有效上覆压力　effective over-burden pressure
有效水头　acting head
有效压力　effective pressure
有效应力　effective stress, final stress
有效应力参数　effective stress parameters
有效应力分析　effective stress analysis
有效应力路径　effective stress path
有效应力原理　principal of effective stress

有效重度　effective unit weight
有效周围压力　effective confining pressure
有斜桩支撑的板桩堤岸　sheet-pile bulkhead with batter pile support
有压流　pressure flow
有压隧洞　tunnel under pressure
有阻尼振动　damped vibration
诱发地震　induced earthquake
诱发应力　induce stress
淤积黏土　warp clay
淤泥　mire, muck
淤泥地基　muck foundation
淤泥质土　mucky soil
淤塞　clogging, blinding, choking up
余震　aftershock, post earthquake
雨水　meteoric water
预测　prognosis
预测值　predicted value
预防性托换　precautionary underpinning
预计沉降量　predicted settlement
预加力　pre-applied force
预加应力法　prestressing method
预剪　preshear
预浸水　prewetting
预裂爆破法　presplit blasting
预裂法　presplitting
预试桩　preliminary test pile, pretest pile
预填骨料混凝土　prepakt concrete
预先抽水法　predraining method
预先开挖　pre-excavation
预压　precompression
预压法　preloading method
预压排水固结　consolidation by preloading and drainage

预压填土	preconstruction fill
预应力撑杆加固法	strengthening method with prestressed brace bar
预应力传递长度	transfer length of prestress
预应力钢筋传力长度	transfer length for pretension tendon
预应力钢缆	strand
预应力混凝土管柱	prestressed concrete drilled caisson, prestressed-concrete cylinder
预应力混凝土桩	prestressed concrete pile
预应力锚杆	prestressed ground anchor bar
预应力损失	loss of prestress, prestressed loss, prestressing loss
预应力损失试验	test for losses of prestress
预应力土锚	prestressed soil anchor
预应力岩石锚杆	prestressed rock anchors
预制钢筋混凝土方桩	precast reinforced concrete square-pile
预制钢筋混凝土管桩	precast reinforced concrete pipe-pile
预制钢筋混凝土实心桩	precast reinforced concrete solid pile
预制钢筋混凝土桩	precast reinforced concrete pile
预制构件拼装结构	sectional construction
预制管水底隧道	prefabricated subaqueous tunnel
预制混凝土板桩	precast concrete sheet pile
预制混凝土桩	precast concrete pile
预制桩	prefabricated pile, pre-formed pile
预注浆	preliminary grouting
预钻孔	preboring, predrilling
元古代	Proterozoic era

原地沉积	deposit in situ
原地面	original ground
原河床	original riverbed
原生黄土	primary loess
原生节理	original joint
原生结核	primary concretion
原生矿物	original mineral, primary mineral
原生黏土	primary clay
原生水	connate water, juvenile water
原生土	genetic soil, residual clay, sedentary soil
原生岩石	rock in place
原始沉积构造	primary sedimentary structure
原始地层压力	original pressure
原始上覆压力	virginal overburden pressure
原始压缩曲线	virgin compression curve
原始应力	virgin stress
原位测试	in-situ test(s)
原位加固土柱	in-situ stabilized column
原位加州承载比试验	in-situ CBR test
原位密度	in-place density
原位土工试验	in-situ soil test
原位推力试验	in-situ thrust test
原位应力	in-situ stress
原位直剪试验	in-situ direct shear test
原型观测	prototype observation
原型监测	prototype monitoring
原型模型	prototype model
原型试验	prototype test
原型钻头	prototype drill
原岩	original rock
原岩应力	stress in original rock
原岩应力比值系数	stress-ratio coefficient of original rock

原岩应力场　initial rock stress field
原状黏土　intact clay
原状试样　intact specimen
原状土　undisturbed soil, undisturbed sample, in-situ soil
原状土试样　sample of undisturbed soil
原子键　atomic bonds
原子键力　interatomic bonding force
圆弧分析法　circular arc analysis
圆弧滑动面　circular sliding surface, circular slip surface
圆弧破坏分析　circular failure analysis
圆孔筛　round-hole sieve
圆筒桩　cylindric(al) pile
圆形基础　circular foundation
圆形围堰　circular cofferdam
圆柱体强度　cylinder strength
圆锥承重试验　cone bearing test
圆锥触探器　conical penetrometer
圆锥触探试验　cone penetration test (CPT)
圆锥触探仪　cone penetrometer
圆锥探头阻力　cone resistance
约束　restrain
约束变形　restraint deformation
约束端点　restrain end
约束反力　constraint reacting force
约束方程　constraint equation
约束节点　restrain joint
约束力矩　restraining moment
约束扭转　restrained torsion
约束扭转常数　restrained torsional constant
约束条件　restraint condition
约束系数　restraint coefficient
约束应力　restraint stress

约束桩　restrained pile
月球土力学　lunar soil mechanics
月球土壤　lunar soil
越岭隧道　watershed tunnel
云母　mica
云母大理石　mica-marble
云母片岩　mica-schist
云母玄武岩　mica-basalt
云母页岩　micaceous shale
允许变形值　allowable value of deformation
允许荷载　allowable load
允许挠度　permissible deflection
允许收缩量　shrinkage allowance
运动黏滞系数　coefficient of kinematic viscosity
运动硬化　kinematic hardening
运动约束　constraint of motion
运积土　transported soil, traveled soil

Z

杂乱土层　erratic subsoil
杂色砂岩　mottled sandstone
杂填土　miscellaneous fill
灾害地质学　disaster geology
载荷　size of load
载荷分量　sharing part of the load
载荷试验　load-bearing test, loading test
载荷台　loading platform
载荷状态　state of loading
再饱和　resaturation
再沉积　redeposition, reprecipitation
再沉积岩　resedimented rock
再处理　retreatment

再分配	redistribution
再固结	reconsolidation
再固结体应变	reconsolidation volumetric strain
再加载	reloading
再压缩	recompaction
再压缩曲线	recompression curve
暂冻土	briefly frozen soil
暂时含水层	temporary water-bearing layer
凿	chisel
凿岩爆破	rock blast
凿岩机	rock boring machine
凿岩效率	rock penetration performance
造岩矿物	rock-forming mineral
噪声污染	nose pollution
增量初应变法	incremental initial strain method
增量初应力法	incremental initial stress method
增量刚度	incremental stiffness
增强充填	reinforcing filler
增强应力	reinforced stress
增塑剂	plasticizer
憎水胶体	hydrophobic colloid
粘结强度	bond strength
粘结土地基	cohesive soil foundation
粘限	sticky limit
詹布法	Janbu method of slope stability analysis
张弛损失	relaxation loss
张断面	tensile fracture
张紧夹具	stress accommodation
张开裂度	gapping fissure
张壳构造	stressed shell construction
张拉成形法	stretch forming
张拉程序	tensioning procedure
张拉钢筋	stretching wire
张拉过程	stressing process, stretching process

张拉计	tensometer
张拉夹具	tension grip
张拉结构	tension structure
张拉试验	pulling test
张拉应变	stretching strain
张拉应力	jacking stress, stretching stress
张拉装置	tensioning device, tensioning equipment
张力	stretching force, tensioning force, tension
张力冲击	tensile impact
张力计	tensiometer
张力曲线	tension curve
张裂缝	tension fissure, tension crack
张裂面	plane of tensile fracture
张裂区	cracked tension zone
胀破强度	burst strength
胀破试验	burst tearing test
沼煤	moor peat
沼泽	bog, marsh, morass
沼泽地	fen land, marshland
沼泽腐殖土	bog muck
沼泽泥炭土	bog peat
沼泽土	fen soil, marsh soil, moor soil, swamp soil
遮帘作用	barrier effect
折减系数	reduced factor
折减因素	factor of reduction
折算弹性模量	reduced elastic modulus
折算高度	reduced height
折算荷载	reduced load
折算密度	reduced density
折算深度	reduced depth
折算水头	reduced head

折算应力	reduced stress
折算值	reduced value
褶皱	fold
褶皱岩石	folded rock
褶皱作用	crumpling
针刺法	needle-punched process
针刺土工织物	needle-punched geotextile
针孔试验	needle hole test, pinhole test
针式贯入仪	needle penetrometer
针织物	knitted fabric
针状构造	needle-shaped structure
针状晶体	acicular crystal
珍珠岩	perlite
真空泵	vacuum pump
真空度	degree of vaccum
真空固结	consolidation by the vacuum method
真空井点	vacuum well point
真空井点系统	vacuum well system
真空取样器	vacuum sampling tubes
真空三轴试验	vacuum triaxial test
真空预压	vacuum preload, vacuum preloading
真空预压法	vacuum method, vacuum preloading method, atmospheric pressure method
真密度	real specific gravity
真内摩擦角	angle of true internal friction, true angle of internal friction
真黏聚力	true cohesion
真三轴试验	true triaxial test
真三轴试验仪	cubical triaxial test apparatus
真三轴仪	true triaxial apparatus
真实断裂应力	actual breaking stress
真实黏度	true viscosity
真相对密度(比重)	true specific gravity

真应变 true strain
真应力 true stress
真值 true value
砧 anvil
振沉桩 vibrator sunk pile
振冲法 vibroflotation method
振冲加固 stabilization by vibroflotation
振冲密实法 vibro-compaction method
振冲器 vibroflot
振冲碎石桩 vibration replacement stone column, vibro-replacement stone column
振冲置换法 vibro-replacement method
振动 flutter
振动拔桩机 vibrating extractor, vibro-driver extractor
振动沉桩 vibrosinking of pile
振动沉桩法 vibrosinking method
振动打锤机 vibrating pile hammer
振动打桩 pile driving by vibration
振动打桩机 vibration pile-driver, vibratory driver, vibro driver, vibro-pile driver
振动法压实 vibration compaction
振动固结，振动压密 consolidation by vibration
振动灌注桩 vibro-pile
振动夯 vibro tamper, vibro-rammer
振动荷载 vibrating load
振动混凝土柱(桩) concrete vibra column
振动碾 vibro-roller
振动碾压法 compaction by vibrating roller
振动平板压实机 vibrating base plate compactor
振动器 vibrator

振动取芯器　vibrocorer
振动取样方法　vibrocoring method
振动三轴仪　vibration triaxial apparatus
振动筛　oscillating screen, pulsating screen, vibration screen
振动深层压密　deep compaction by vibration
振动台　jigging platform, percussion table
振动循环　cycle of vibration
振动压实　compaction by vibration, vibratory compaction
振动压实机　vibrating compactor
振动液化试验　vibratory liquidation test
振动置换　vibro-replacement
振动周期　period of vibration
振动桩锤　vibratory pile hammer
振动钻井　vibro drilling
振幅　amplitude of vibration
振幅比　amplitude ratio
振幅放大因素　amplitude magnification factor
振密挤密法　vibro-densification method
振实理论　theory of vibrating densification
振松　decompaction
振型　mode of vibration
震动分析　seismic analysis
震动加速度　seismic acceleration
震动台试验　shaking table test
震动效应　seismic action effect
震害　earthquake damage, earthquake hazards
震级　magnitude
震级测定　magnitude determination
震级分布　magnitude distribution
震级图　magnitude chart
震裂岩石　shatter rock
震陷　earthquake subsidence, seismic subsid-

ence
震源　earthquake focus, earthquake hypocenter, hypocenter
震中　earthquake center, epicentre
震中对点　anticenter of earthquake, anti-epicenter, anti-epicentrum
震中距　epicentral distance
震中烈度　epicentral intensity
蒸发　evaporation
蒸发总量　evapo-transpiration
蒸馏水　distilled water
蒸汽渗透阻　resistance to water vapor permeability
整合地层　conformable stratum
整体刚度　integral rigidity
整体滑动　complete sliding
整体剪切破坏　general shear failure, general-shear failure
整体破坏　block failure
整体倾斜　overall inclination
整体屈曲　overall buckling
整体取芯法　integral coring method
整体失稳　overall unstability
整体式挡土墙　monolithic retaining walls
整体特性　mass properties
整体弯曲　overall flexure
整体稳定　overall stability
整体移动　mass movement
正变质岩　orthometamorphite
正长石　orthoclase
正长岩　syenite
正常固结土　normally consolidated soil
正常固结线　normal consolidation line
正常使用极限状态　serviceability limit state

正常使用阶段　serviceability stage
正割弹性模量　secant modulus of elasticity
正割刚度法　secant method
正割曲线　secant curve
正割屈服应力　secant yield stress
正割圆　secant circle
正交测斜灵敏度　cross sensitivity
正交各向异性　orthotropy
正交各向异性材料　orthotropic material
正交各向异性弹性体　orthotropic elasticity
正交各向异性的　orthotropic
正交各向异性颗粒　orthotropic particle
正交条件　normality condition
正切曲线　tangent curve
正态分布　normal distribution
正态分布律　normality law
正向铲　face shovel
正循环冲洗　direct flushing
正循环旋转钻进　normal-circulation rotary drilling
正应力　direct stress
支撑　bracing, propping
支撑挡板　braced sheeting
支撑垫板　support ped
支撑工程　shoring work
支撑开挖　braced excavation
支承桩　bearing pile
支顶加固　strengthening by struts
支墩式挡墙　buttress retaining wall
支护桩　tangent pile
支柱基础　stanchion base
支座下沉　settling of support
织物模板　fabric forms
直方图　histogram

直剪试验　direct shear test
直剪仪　direct shear apparatus
直接单剪试验　direct simple shear test [DSS-test]
植草护坡　seeded slope
植树护坡　vegetation cover of slope protection
止冲流速　non-scouring velocity
止水　shut-off water, water stopping
止水层　seal coat
止水带　water stopping band
止水缝　sealed joint
止水胶垫　water stopping rubber gasket
止水片　waterstop blade
止水墙　seep proof screen
止水设施　seal installation
止水帷幕　water tight screen, watertight screen
纸板排水　cardboard drain, paper drain
指标特性试验　index property test
指定贯入度　specified penetration
趾墙　toe wall
制桩场　pile fabricating yard
质点动力学　particle dynamics
质量比　mass ratio
质量-弹簧-阻尼器体系　mass-spring-dashpot system
质量控制　quality control
质量力　body force
质量密度　mass density
质子磁力仪　proton magnetometer
致密裂缝　tight fissure
致密岩石　compact rock
智能灌浆参数记录仪　intelligent grouting parameter recorder
滞后　retardation

滞回环　hysteresis loop
滞回模量　hysteresis modulus
滞回压实　hysteresis compaction
蛭石　roseite
置换　substitution
置换地基　replaced ground base
置换法　replacement method
置换基础　replaced foundation
置换挤密　compaction by compaction
置换率　displacement ratio, replacement ratio
置换柱　replacement pile
置信度　confindence level
置信界限　confindence limit
中层支撑　secondary story in tunnel timbering
中等稠度　medium consistency
中等岩石条件　average rock condition
中级颗粒　intermediate grain
中间试验　pilot test
中粒土　medium grained soil
中溶盐试验　moderately soluble salt test
中砂　medium sand
中生代　Mesozoic era
中线法　centering method
中心荷载　concentric load, central load
中心线　center line
中心震　central earthquake
中心桩　center peg
中性土壤　neutral soil
中性压力　neutral pressure
中性岩　intermediate rock
中性应力　neutral stress
中央航空局土分类法　federal aviation agency classification [FAAC]
中值粒径　median(medium)diameter

中主平面　intermediate principle plane
中主应变　intermediate principle strain
中主应力　intermediate principle stress
中子测井　neutron logging
终点压力　terminal pressure
重锤夯实　heavy tamping, ramming by heavy hammer
重度　unit weight
重复荷载　repeated loading
重复试验　repeated-load test
重结晶作用　recrystallization
重晶石　barite, barytes, heavy spar
重矿物　heavy minerals
重力锤　gravity hammer
重力断层　gravity fault
重力加速度　acceleration of gravity
重力勘探　gravitational exploration
重力排水　drainage by gravity
重力式挡土墙　gravity retaining wall
重力势　gravitational potential
重力水　gravitational water
重力仪　gravimeter
重力作用　action of gravity
重量百分比　percentage by weight
重量损失　loss of weight
重黏土　heavy soil
重塑度　degree of remoulding
重塑黏土　remolded clay
重塑强度　remolded strength
重塑土　remolded soil
重塑土特性　remolded properties
重塑性　remoldability
重塑指数　remolding
重心　centroid

重型基础	heavy foundation
重粘土	heavy textured soil
重整坡度	regrade
周边灌浆	perimeter grouting
周边效应	border effect
周期单剪试验	cyclic simple shear test
周期荷载	cyclic loading, cyclic load, periodic load
周期加荷三轴试验	cyclic triaxial test
周期加热试验	cyclic heat test
周期抗剪强度	cyclic shear resistance
周期衰化	cyclic degradation
周期应变软化	cyclic strain softening
周期应力	cyclic stress
周期载荷试验	cyclic load test
周期振动	periodical vibration
周围压力	all round pressure, ambient pressure
轴对称	axial symmetry
轴对称固结	axially symmetric consolidation
轴对称应力	axisymmetric stress
轴线放样	setting out of axis
轴向变形	axial deformation
轴向承载力	axial bearing capacity
轴向荷载	axial load
轴向荷载试验	axial load test
轴向减压断裂试验	decreasing axial pressure fracture test
轴向拉伸试验	axial extension test
轴向力影响系数	influence factor for axial force
轴向压力	axial compressive force
轴向压缩	axial compression
轴向应变	axial strain

轴向应力 axial stress
轴心受力构件强度 strength of axially-loaded member
侏罗纪 Jurassic period
烛煤页岩 cannel shale
主动滑动面 active surface of sliding
主动滑动区 active slide area
主动剪应力 active shear stress
主动朗肯区 active Rankine zone
主动力 active force
主动塑性平衡状态 active state of plastic equilibrium
主动土压力 active earth pressure
主动土压力系数 coefficient of active earth pressure
主动抑压法 active containment alternative
主动桩 active pile
主拱圈 main arch ring
主固结 primary consolidation
主价键 primary valence bond
主剪应力 principal shearing stress
主平面 principal plane
主蠕变 primary creep
主时间效应 primary time effect
主体岩石 host rock
主压应力 principal compressive stress
主要缺陷 major defect
主应变 principal strain
主应变轴 principal axes of strain
主应力 principal stress
主应力比 principal stress ratio, ratio of pricipal stress
主应力法 principal stress method
主应力迹线 isostatic

主应力空间　principal stress space
主应力圆　principal stress circle
主应力轴　principal axes of stress
主震　principal earthquake
主桩　key pile
注浆法　injection process
注水井　injection well, input well
注水试验　water injecting test
注水压实　compaction by watering
贮槽　sump
柱脚　pedestal
柱式边墙衬砌　lining with column-typed sidewalls
柱下钢筋混凝土条形基础
　　reinforced concrete strip foundation under columns
柱桩　column pile
柱状结构　columnar structure
柱状图　columnar section
筑坝　damming
筑岛沉井　sinking of open caisson on filled up island
铸钢井壁　steel casing
抓斗挖土机　grabbing excavator
抓拉试验　grab tensile test
砖红壤性土　lateritic soil
砖基础　brick foundation
转动惯量　rotary inertial
桩　pile
桩侧摩擦力　side friction resistance of pile, skin friction of pile
桩侧向荷载试验　lateral pile load test
桩侧阻　shaft resistance
桩承筏基　pile supported raft

桩承基础	pile supported footing
桩承基脚	pile footing
桩承台	pile grid, platform of piles
桩承台梁	pile capping beam
桩承作用	pile action
桩锤	driving hammer, pile hammer
桩锤垫	hammer cushion
桩的布置	arrangement of piles
桩的颤动	pile flutter
桩的承载力	pile capacity
桩的冲孔	jetting of pile
桩的动荷载试验	dynamic load test of pile
桩的腐蚀	pile corrosion
桩的负摩阻力	negative skin friction of pile
桩的贯入	pile penetration
桩的贯入度	set of pile
桩的荷载传递函数法	load transfer function method of pile
桩的荷载试验	vertical loading test of pile
桩的横向承载力	lateral bearing capacity of piles
桩的横向载荷试验	lateral loading test of pile
桩的回弹	rebound of pile
桩的极限荷载	ultimate load of pile
桩的截面	pile cross section
桩的静力公式	static formula of pile
桩的抗拔试验	pulling test on pile, uplift load test of pile
桩的隆起	pile heave
桩的梅花式排列	staggered arrangement of piles
桩的拼接	pile splice
桩的深度-阻力曲线	depth resistance curve of pile

桩的竖向反力系数	coefficient of vertical pile reaction
桩的水平向反力系数	coefficient of horizontal pile reaction
桩的套护	pile jacketing
桩的完整性试验	pile integrity test
桩的行列式排列	arrangement of piles in rank form
桩的载荷试验	load test on pile
桩的制作	pile manufacture
桩的中性点	neutral point of pile
桩的轴向承载力	axial bearing capacity of piles
桩的轴向容许承载力	axial allowable bearing capacity of pile
桩的最小间距	minimum pile spacing
桩垫	dolly, pile cushion
桩顶连系梁	pile cap beam
桩顶约束条件	restrain condition of pile top
桩端	point of pile
桩端承力	point-bearing capacity
桩端承载力	pile end bearing
桩端承载能力	end-bearing capacity of pile
桩端阻力	end resistance of pile
桩对承台的冲切	punching of pile on cap
桩负摩阻处理	treatment of negative skin friction
桩复打试验	pile retapping test
桩工机械	pile driving machinery
桩箍	lagging of pile, pile band, pile hoop, driving band, drive
桩基础	pile foundation
桩基的动力试验	dynamic test of pile(s)
桩基试验	pile test
桩架	ram guide

桩架平台	platform of pile frame
桩尖	pile point, pile tip, pile toe
桩尖阻力	tip resistance
桩截水墙	pile cut-off
桩静载试验	pile loading test
桩距	pile spacing
桩力计	pile force gauge
桩锚	pile anchor
桩帽	capblock, driving cap, driving helmet, end cap, helmet, pile cap, cap of pile
桩帽加固法	pile cap method
桩排架	pile bent
桩墙	piling wall
桩群	cluster, pile cluster
桩群作用	group action
桩身	pile body, pile shaft
桩式靠船建筑	dolphin type breasting structure
桩式码头	piled wharf
桩式桥墩	pile pier
桩式桥台	pile bent pier
桩式托换	piling underpinning
桩式围堰	pile cofferdam
桩束(群)	clustered piles
桩数	number of piles
桩头	butt, pile head
桩—土荷载比	pile-soil loading ratio
桩—土相互作用	pile-soil interaction
桩—土应力比	pile-soil stress ratio
桩位布置	pile layout(plan)
桩芯	pile core
桩靴	driving shoe, pile shoe, drive shoe
桩压入法	pile jacking method
桩支承	pile bearing
桩止点	pile stoppage point

装有量测元件的桩　instrumented pile
装载机　loader
状态变量　state variable
状态参数　state parameters
锥击试验　cone penetration impact test
锥孔　ream
锥式液限仪　cone penetrometer for liquid limit test
锥形基础　cone foundation
锥形桩　conical pile, tapered pile
准饱和土　quasi-saturated soil
准超固结　quasi-overconsolidation
准弹性定律　hypoelastic law
准固结压力　pseudo-consolidation pressure
准静力触探贯入度　quasi-static cone penetration
准静力法　pseudo-static approach, quasi-static method
准黏聚力　pseudo-cohesion
准三向固结理论　pseudo-three-dimensional consolidation theory
准先期固结压力　quasi-preconsolidation pressure
准岩体强度　pseudo-strength of rock mass
准永久值　quasi-permanent value
准永久组合　quasi-permanent combination
卓越周期　predominant period
自承式静压桩　self-supporting static pressed pile
自动冲锤打桩机　automatic ram pile driver
自动调节集管　automated header
自动灌浆设备　automated grouting equipment
自动制浆站　automated grout plant
自动注入系统　automated injection system

自流承压水 flowing artesian
自流井 artesian well, gusher type well
自流水压力 artesian pressure, artesian head
自录计程仪 autographic odometer
自落填缝反滤层 collapsible filter
自然保护 natural conservation
自然边界条件 natural boundary conditions
自然沉淀 natural sedimentation
自然干燥 natural drying
自然环境 natural environment
自然浸渍 natural impregnation
自然侵蚀 natural erosion
自然休止角 angle of nature repose
自升式钻架 jack-up rig
自压实性 ability to self-compact
自应力法 self-stressing method
自应力水泥 self-cement for stressing
自由度 degree of freedom
自由端三轴试验 free-end triaxial test
自由高度 unsupported height
自由膨胀率 free swelling ratio, rate of free expansion
自由膨胀率试验 free swell test
自由水 free water
自由水位 free water elevation
自由振动 free vibration
自由周期 free period
自振柱试验 free vibration column test
自重固结 self-consolidation
自重湿陷系数 coefficient of self-weight collapsibility
自重湿陷性黄土 self weight collapse loess
自重应力 gravity stress, self-weight stress, geostatic stress

自钻式旁压仪　self-boring pressure meter
综合工程地质图　comprehensive engineering geological map
综合利用　multiple use
总变形　total deflection
总变形量　total deformation
总沉降量　total settlement
总尺寸　overall dimension
总管　header pipe
总含水量　total moisture content
总隆起量　total heave
总容许荷载　allowable gross bearing pressure
总容许限度　total tolerance
总湿陷量　total collapse
总水灰比　total water cement ratio
总位移量　total displacement
总应力　overall stress, total stress
总应力分析　total stress analysis
总应力路径　total stress path
总应力破坏包线　total stress failure envelope
总应力强度参数　total stress strength parameter
总重量　total weight
总桩端阻力　point resistance force
总钻深　total depth
纵波　longitudinal wave
纵剖面　longitudinal profile
纵向刚度　longitudinal rigidity
纵向裂缝　longitudinal crack
纵向透水性　longitudinal permeability
纵向弯曲　longitudinal buckling
纵向弯曲系数　longitudinal bending factor
纵向位移　longitudinal displacement
走滑断层　strike-slip fault
走向　strike direction

足尺试验　full scale test
阻抗比　impedance ratio
阻抗系数　resistance factor
阻力曲线　resistivity curve
阻沫剂　antifoam
阻尼　damping
阻尼比　damping ratio
阻尼矩阵　damping matrix
阻尼理论　theory of damping
阻尼器　dashpot
阻尼因素　damping factor
阻水层　aquiclude
组构　fabric
组构分析　fabric analysis
组构各向异性　fabric anisotropy
组合型土工织物　composite geotextile
钻爆参数　parameter of drilling and blasting
钻采平台　drilling platform
钻杆　bore rod, boring rod, drilling rod, rod
钻杆长度　length of drill rod
钻杆导向器　screw holder
钻杆转速　rod rotation speed
钻工　borer
钻管　drill pipe
钻机　boring machine, drilling rig
钻架　boring frame, rock drill mounting
钻进转速　rotation speed for drilling
钻孔　bore hole, drill hole, sinking of bore hole
钻孔泵　bust pump
钻孔沉井　drilled caisson
钻孔电视　borehole TV
钻孔堵塞　bridge
钻孔墩　drilled pier

钻孔灌注桩	bored cast-in-place pile, drive cast-place-pile, non-displacement pile
钻孔规	borehole caliper
钻孔回填	plugging
钻孔记录	borehole log, boring log, borehole record, log of bore hole
钻孔剪切仪	borehole shear apparatus
钻孔检查显示器	borescope
钻孔勘探	borehole surveying
钻孔扩底灌注桩	under-reamed bored pile
钻孔泥浆	drilling mud
钻孔千斤顶	borehole jack
钻孔潜望镜	borehole periscope
钻孔日报表	daily drilling report
钻孔扫描器	borehole scanner
钻孔竖井	drilled shaft
钻孔土样	bore plug, borehole sample, borehole specimen
钻孔压力恢复试验	borehole pressure recovery test
钻孔岩芯	borehole core
钻孔异物	fish
钻孔引伸仪	borehole extensometer
钻孔照相机	borehole camera
钻孔直剪仪	borehole direct shear device
钻孔重力仪	borehole gravimeter
钻孔柱状图	bore log, drill log, log
钻孔桩	bored pile, drilled shaft
钻孔最大直径	drilled maximum diameter
钻取岩芯	coring
钻时记录	drilling time log
钻探	boring, drilling, exploration drilling
钻探报告	drilling report
钻探机具	boring rig, drill rig

钻探解释剖面　interpretative log
钻探设备　drilling equipments
钻探绳索采样系统　drilling and wireline coring
钻探套管　boring casting
钻头　boring bit, drill bit, jack bit, bit
钻头规格　gauge of bit
钻斜孔法　off-angle drilling
钻屑　cuttings of boring
钻芯法实验　test by coring sample
钻液　drilling fluid
最大拔桩力　maximum extracting force on pile
最大饱和度　maximum saturation
最大成孔深度　maximum drilling depth
最大冻土深度　maximum depth of frozen ground
最大分子吸水量　maximum molecular water content
最大干密度　maximum dry density
最大干重度　maximum dry unit weight
最大固结压力　maximum consolidation pressure
最大剪应力理论　maximum shear theory
最大孔隙比　maximum void ratio, void ratio in loosest state
最大粒度　limiting grain size
最大粒径　maximum particle size
最大前期压力　maximum past pressure
最大容许荷载　maximum permissible load
最大位移　maximum displacement
最大压桩力　maximum pressure on pile
最大主应力　maximum principal stress
最低冲刷线　lowest erosion line
最佳逼近　optimum approximation
最佳覆盖厚度　optimum depth

最佳含砂率　optimum fine aggregate perentage
最佳级配　optimal gradation
最佳密度　optimum density
最佳压实深度　optimal rolling depth
最小安全稳定性　minimum safety stability
最小二乘法　least-square method
最小贯入度　minimum penetration
最小孔隙比　minimum void ratio, void ratio in densest state
最小势能原理　principal of minimum potential energy
最小余能原理　principal of minimum complementary energy
最优含水量　optimum moisture content, optimurn water content
最优加筋层数　optimum reinforcement layer number
最终沉降(量)　final settlement, ultimate settlement
最终贯入度　final penetration, final set
最终强度　final strength
作用点　point of action
作用分项系数　partial(safety) factor for action
作用应力　applied stress
座板　seat board
座撑　seat stay
座角钢　seat angle
座梁　seat beam